D1716597

THE MICROPROCESSOR PERIPHERAL IC REFERENCE MANUAL

Richard J. Prestopnik
Professor of Electrical and Computer Technology
Fulton-Montgomery Community College

Prentice Hall
Englewood Cliffs, New Jersey 07632

Library of Congress Cataloging-in-Publication Data

PRESTOPNIK, RICHARD J.
The microprocessor peripheral IC reference manual.

Bibliography: p.
Includes index.
1. Integrated circuits—Handbooks, manuals, etc.
I. Title.
TK7874.P734 1989 621.395 88-19557
ISBN 0-13-580705-0

Editorial/production supervision: BARBARA MARTTINE
Cover design: GEORGE CORNELL
Manufacturing buyer: MARY ANN GLORIANDE

The publisher offer discounts on this book when ordered in bulk quantities. For more information, write:

Special Sales/College Marketing
College Technical and Reference Division
Prentice Hall
Englewood Cliffs, New Jersey 07632

Printed in the United States of America

10 9 8 7 6 5 4 3 2 1

ISBN 0-13-580705-0

Prentice-Hall International (UK) Limited, *London*
Prentice-Hall of Australia Pty. Limited, *Sydney*
Prentice-Hall Canada, Inc., *Toronto*
Prentice-Hall Hispanoamericana, S.A., *Mexico*
Prentice-Hall of India Private Limited, *New Delhi*
Prentice-Hall of Japan, Inc., *Tokyo*
Simon & Schuster Asia Pte. Ltd., *Singapore*
Editora Prentice-Hall do Brasil, Ltda., *Rio de Janeiro*

Dedicated, with all the love that a son can give,
to my mother and father, Frances and John Prestopnik.

Contents

Acknowledgments

A number of major integrated circuit manufacturers provided invaluable information used in the production of this book. I am grateful for materials and support from the following companies:

Advanced Micro Devices, Inc.
Analog Devices
GE Solid State
Gould Semiconductors
Intel Corporation
Precision Monolithics, Inc.
Signetics Corporation
Texas Instruments
Zilog

An undertaking of this magnitude cannot be realized without the love, support,and understanding displayed by family members. To my children, Nathan, Emily, and Adam, thank you for the patience you had when Dad was busy at the computer. Whenever I look back to the work involved in writing this book, I'll remember the happy sounds you made playing in the next room.

To my wife, Jan, all my love and appreciation for being a wonderful wife, a terrific mother, and the best proofreader a guy could hope to marry. Any literary style evident within this book is a direct result of her own excellent writing skills and ability to convert technical drivel into understandable English.

Richard J. Prestopnik

How This Book Will Help You

VLSI technology has progressed rapidly within the past few years. Circuits designed only several years ago, which required a significant number of integrated circuits for implementation, can now be implemented by as little as one chip because of the many new ICs available. The chip designers and manufacturers have become so clever that it has become difficult for the design engineer, student, teacher, and industry in general to keep up with what is new and potentially useful. If you have ever been in the position of designing with these new ICs, there were probably several things you wanted to know to guide you in making the proper choices. Questions frequently asked include:

1. *What chip functions are available in the first place?* With so many manufacturers producing these specialty chips, a good deal of time can be spent trying to find out what chips are actually in the marketplace. A good amount of effort is needed just to acquire the manufacturers' data books and other preliminary information.
2. *Will the chip meet the required design needs?* This information is available in the manufacturers' data books once you get past the many timing diagrams and sentences that seem to be made up exclusively of words like "tDW," "tRWVCL," and "tRxRDY CLEAR." It's not that this information isn't critically important, but it does confuse the issue when only the most basic information is needed. Many initial design decisions can be made if one knows whether or not the chip is functionally compatible with the system to be designed. Often the most fundamental of specifications, such as voltage levels, power dissipation, and microprocessor compatibility, can eliminate a chip from further consideration.
3. *Where can the chip be obtained? Is there more than one source?* Once a chip is selected, adequate supplies of the part need to

be available throughout the production life of the design. An alternate source will insure parts availability.

4. *What is the pinout?* Basic physical information about the part, such as the pinout, power dissipation, and temperature requirements, is necessary for making a sound design decision.
5. *What are the basic operating principles?* Once again, the manufacturers' data books give this information, but the reader often has to sift through more material than is necessary just to understand the basic operation of the part. This is not a minor problem. Some of the chips are so complex that basic operating information is not easily gleaned from the data manuals. A functional description with examples is needed to promote quick understanding of the IC in question.

Most designers would probably agree that a reference source that answers all the above questions would be extremely useful. This text is that source. This manual pulls together the critical information needed for a designer, student, or instructor to make a correct choice when selecting an integrated circuit. A primary objective of this reference manual is to set forth the fundamental functional information necessary to understand how a chip works, how it is programmed, and how it is controlled. Therefore, much of the intricate specification detail found in the manufacturers' data books becomes secondary to explanations of the fundamental principles of operation. Once a chip is understood and selected, the data book is the next logical step in the design effort. This approach is recommended to expedite the design and learning effort. The manufacturer's data manual may contain all the currently available information on a particular integrated circuit, but more often than not, the presentation of the detailed information blocks the way to understanding the basic function and operation of the device. Numerous details concerning fine points, such as the chip's intricate time delays, simply clutter the reader's first impression of a new device and bog down the assimilation of fundamental details. The fundamental functional concepts that could ultimately add to the designer's expertise and have a significant bearing on future designs may be ignored by the designer out of frustration or boredom. This is a problem for many engineers and students, often causing them to stay with the tried-and-true methods of design learned from past experience. Not only is this inefficient, from an economic point of view, but it also inhibits the learning and growth process that is crucial to successful engineering.

Features of this manual, which make it an invaluable aid to the professional, include: schematic diagrams of each chip, including pin-out and pin function; detailed tables giving the functions of each pin; and a reference section that organizes chip functions to aid in the selection process. As a further help, all sources for the chips are provided.

For the student, this text should be an eye-opener and reference for future studies and design because it demonstrates what large-scale integration can offer them beyond the flip-flop, counters and shift registers to which they are accustomed. Circuit designers will find this text a useful first step toward finding the right chip for design.

The chips covered in this manual are those that support microprocessor system designs. Such chips include I/O devices, controllers, and timers. This book will not easily go out-of-date since the advanced computer concepts embodied here are likely to remain with us in the foreseeable future. For example, the MK68901 Multi-Functional Peripheral is a support chip for 68000 family microprocessors. It includes I/O ports, an interrupt controller, various timers, and a USART, all on a single piece of silicon. These same functional units are discussed within this book as specific integrated circuits. It is worth noting that even with the rapid growth of VLSI design, the primary logical building blocks in computing systems remain intact. They are merely lumped together on the same chip. The same rationale can be used for chips which control the operation of storage devices, such as floppy disk controllers and hard disk controllers. Because of the large body of design work and economic effort already supporting them, these storage technologies are not likely to disappear in the near future. Insight into the current state-of-the-art is mandatory if an engineer is to assess the evolutionary changes inevitable in this industry. An understanding of these critical building blocks, as discussed in this text, provides the doorway to a fundamental base of knowledge for future IC designs.

Since this text deals with higher-level integrated circuits, some assumptions are made regarding the reader's background. To make the best use of this manual, basic knowledge in the following related areas is necessary: an understanding of the binary and hexadecimal systems, an understanding of digital logic, and a basic knowledge of microcomputer terminology. The manual is written to facilitate an understanding of the ICs presented and is not written assuming a large base of experience in the areas mentioned. A preliminary section covering basic microprocessor fundamentals is included.

To further enhance the value of this text, a directory of integrated circuits is provided. The sheer number of integrated circuits available makes it impossible to report on each one. This directory provides basic "fingertip" information on all available related integrated circuits. This information includes part number, function, and manufacturers. Rounding out the book are a bibliography of useful books and reference manuals, and a compilation of the components in current microprocessor families.

A large portion of this book is devoted to detailed, functional explanations of representative peripheral chips. These will enable the reader to:

1. Understand the functional operation of the chip, including details on overall organization, programming, and CPU interfacing.
2. Gain insight into many of the concepts used to produce a sophisticated computer system. The discussion of the chips provides an introductory tutorial into many areas, including Direct Memory Access (DMA), Interrupts, and Communications.
3. Become familiar with VLSI circuits and the implementation philosophy surrounding them. An understanding of these chips will enable the interested reader to see how a single chip, with a limited number of physical pins, has been designed to provide maximum capability.

SECTION I

A Microprocessor Primer

The microprocessor chip (*microprocessing unit—MPU, or central processing unit—CPU*) is the heart of many personal computers, consumer products, and other digital circuits. Microprocessors are generally only a part of what is known as a *microprocessor family.* A microprocessor family typically includes several versions of the microprocessor tailored for specific performance applications (i.e., addressable memory, speed, programmability). Additionally, a host of peripheral chips designed to allow an easy and efficient expansion of a microprocessor-based system is common in microprocessor families. The peripheral integrated circuits discussed in this book are used to augment microprocessor-based systems. A workable microprocessor system can easily be assembled from a basic microprocessor, memory, and a small amount of additional hardware. This rudimentary system can be transformed, through straightforward enhancements using some of the most basic peripheral chips available, into a sophisticated system with computing power and high performance. This section of the book sets the background and fundamental principles under which the microprocessor and peripheral chip must operate.

INFORMATION EXCHANGE ALONG THE DATA BUS, ADDRESS BUS, AND OTHER PATHWAYS

Information between the CPU and other components in the microcomputer system is exchanged by means of the *data bus.* Figure I.1 illustrates the block diagram of a typical microprocessor system. The data bus is *bidirectional;* data can be transferred in either direction along the bus. The data bus is also a three-state (high, low, high-impedance) bus since it is shared among many system components. The use of a three-state (also known as tri-state) data bus allows the bus to be used by more than one device at a time. Devices not using the bus are forced into a tri-state condition and do not appear to be attached to the bus from an electrical point of view. This eliminates the interference and bus contention that would occur if two or more devices tried to access the bus at the same time.

The flow of data along the data bus occurs in conjunction with addressing information present on the *address bus.* The addressing information gives direction to the data information. Most devices attached to the data bus are referenced by an address. The software program executing within the CPU produces the appropriate addresses so that data comes from or goes to the proper place. The address bus is usually a tri-state, unidirectional bus.

Figure I.1 Microprocessor System Block Diagram

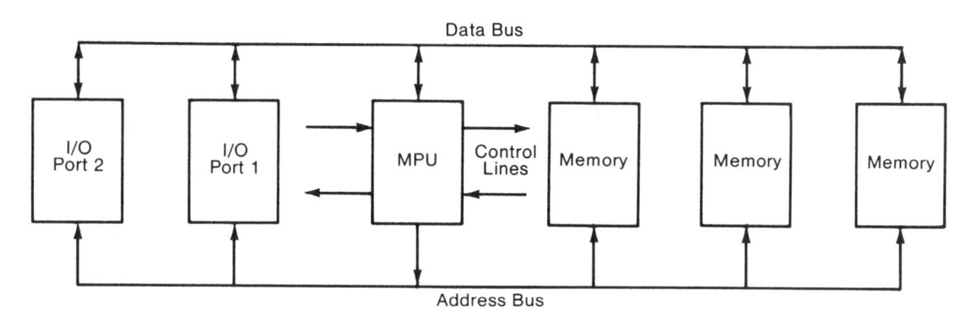

A data transfer actually occurs under the guidance of the *control bus.* The control bus is not as well defined as the address or data bus. Typically, it consists of a loose grouping of the CPU control lines that synchronize the exchange of data among system components. Read, Write, or S0 are typical names for control bus signal lines.

The size of the data bus is usually indicative of the performance classification of a microprocessor as well as a defining element for the types of peripherals to which it will be attached. Common data bus sizes are 4-, 8-, 16-, and 32-bits. This means, for example, that data is transferred in quantities of 4, 8, 16, or 32 bits. Generally, the larger the data bus, the greater the throughput in a system. Other factors, however, can affect system performance as well. Data bus sizes other than those mentioned are available, but are not as common, nor are they considered to be standard bus sizes.

The size of the address bus determines the amount of directly addressable memory in a microcomputer system. This implies that the *instruction set* (software programming instructions) for a particular microprocessor is geared toward a specific maximum memory size. (The upper boundary on memory size is often overcome by programming techniques, such as bank addressing, in which additional memory can be added and utilized in a microprocessor system.) Each additional bit in the address bus will double the size of the existing memory space. For instance, a 16-bit address bus is capable of addressing 65,536 addresses. If a single, additional address bit could be added to the bus, 131,072 addresses would be possible. Common address bus sizes are 16, 20, and 24 bits.

One limiting factor in integrated circuit design is the number of pins that can be physically attached to the chip. *Multiplexing* techniques are used to make better use of the available pins. The address bus and data bus are often multiplexed. An example of this multiplexed technique is used in the 8085 microprocessor (Fig. I.2a). The address bus is 16-bits long; the data bus is 8-bits long. The eight data pins share the address pins labelled ADO - AD7. During normal CPU processing, address and data information are not present on the multiplexed portion of the bus at the same time. At varying points in time, addressing information and then data information are present on the multiplexed bus. Some external hardware is necessary to retain this transient information for later processing. Peripheral chips, designed to match a particular microprocessor family, often are designed with this hardware. Contrast the 8085 with its predecessor, the 8080 (Fig. I.2b), which does not use a multiplexed bus. The 8080 uses 24 pins to contain the same address and data capability that the 8085 does in 16 pins. Due to the use of multiplexing, some of the resulting extra pins on the microprocessor can be used to provide other, more powerful functions.

The internal timings of the microprocessor are derived from a *system clock*. To insure accuracy and precision, a crystal oscillator is usually employed as the primary timing element. Most of the oscillator circuitry is included within the microprocessor or an associated clock generator chip. The oscillator is completed by the addition of the crystal and several stabilizing capacitors. The system clock frequency is obtained from the output of internal frequency divider circuits so that it is lower than the oscillator frequency. Most of the microprocessor

Figure I.2 Differing Microprocessor Bus Techniques

Courtesy of Intel Corp.

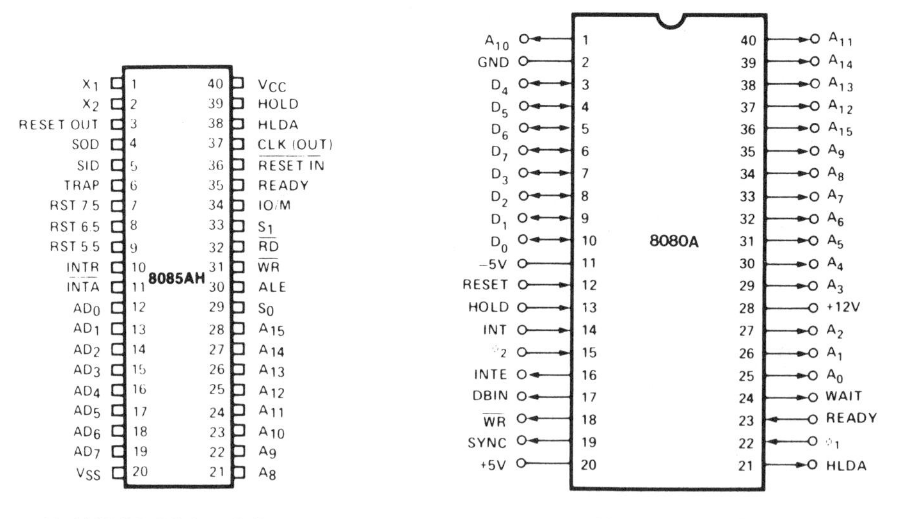

(a) 8085 Multiplexed System **(b) 8080 Unmultiplexed**

logic is now synchronized with the heartbeat of the system clock. Many microprocessors are available in varying performance classes that are largely related to those system clock frequencies that they can handle. For instance, the Z80 is available in 8 MHz, 6 MHz, 4 MHz, and 2.5 MHz versions. Since all cycle times will be controlled by the system clock, a fast clock implies faster execution of software instructions.

MICROPROCESSOR REGISTERS—HOW THEY CONTROL SYSTEM OPERATION

The size and number of CPU registers vary depending on the microprocessor chosen. Most microprocessors have several general purpose registers used to temporarily hold data or addresses. The bit size of these registers is usually the same as the data bus size. Therefore, an 8-bit machine would have 8-bit registers. In some cases, a general purpose register can be grouped with another register to form a register pair. This capability extends the bit size of the register and is useful for address storage or mathematical operations. One of the general purpose registers is often referred to as the *accumulator.* This register is used extensively for arithmetic, I/O, and data transfer operations.

The *program counter* contains the address of the next instruction for execution. For the majority of software instructions, the program counter address will automatically increment by one, allowing the processor to point to the next instruction in memory (instructions are stored sequentially). The program counter can also change in a nonsequential fashion, such as in the case of a branching or subroutine call instruction. In this instance, a new address will overwrite the address in the program counter, thus pointing the processor to a new set of instructions in another memory area. Larger and more powerful processors may have *segment registers* which will allocate large blocks of memory to specific uses.

Instructions procured from memory are transported across the data bus to the CPU and latched within the *instruction register* for decoding. The results of the decode process will determine the next sequence of events performed by the processor.

One other useful register in most microprocessors is the *stack pointer.* The stack pointer holds the address of a portion of memory set aside for a stack area. Some operations, such as subroutines and interrupts are best implemented with a stack structure. Subroutines and interrupts alter the normal process of execution for a period of time. Under these circumstances, the address of the next normal instruction must be saved, as must the current register contents. This mechanism lets the programmer *push* addresses and data onto the stack for temporary storage. The CPU can be restored to its previous state by *popping* this information off the stack. A typical programming process would have data and addresses taken from the stack in reverse order from that in which they were placed there (First In, Last Out - FILO). Thus, the internal data and address registers are reestablished to their original values as if the subroutine or interrupt never took place.

Understanding Read and Write Operations

Two fundamental processes characterize the bulk of CPU operations. They are the *read* and *write* operations. Even seemingly complex CPU procedures can be viewed as nothing more than simple reads and writes. Since the CPU dictates the course of events in the computer system, all actions are with respect to the CPU. That is, a read operation means that the CPU will acquire data from some portion of the computer system, such as memory or an input port. A write operation means that the CPU will present data to some portion of the computer system, such as memory or an output port. The following section explains how the processor initiates read and write operations.

Understanding Fetch and Execute Cycles

The microprocessor operates in a repetitive fashion. Instructions are obtained from memory, referred to as an *instruction fetch cycle,* and then executed. This process is cyclic in nature (Fig. I.3). Assume that an instruction that will move data from a memory location to an internal register (read operation) is about to be processed. During the instruction fetch, the processor places onto the address bus the memory address where the instruction is stored. Address decoding circuits, external to the microprocessor, will select the actual memory chip containing this instruction. This selection process, along with appropriate control bus signals, will cause the memory chip to place its addressed contents onto the data bus. This data is then placed into the microprocessor's instruction register for decoding.

The results from the decoding of the instruction in this example will initiate an *execute cycle* similar to that of the instruction fetch. Execution means that the instruction, now known to the CPU, is acted upon. The address of the data (assumed to be residing within the microprocessor) is placed onto the address bus which in turn causes a

Figure I.3 Instruction Fetch Cycle

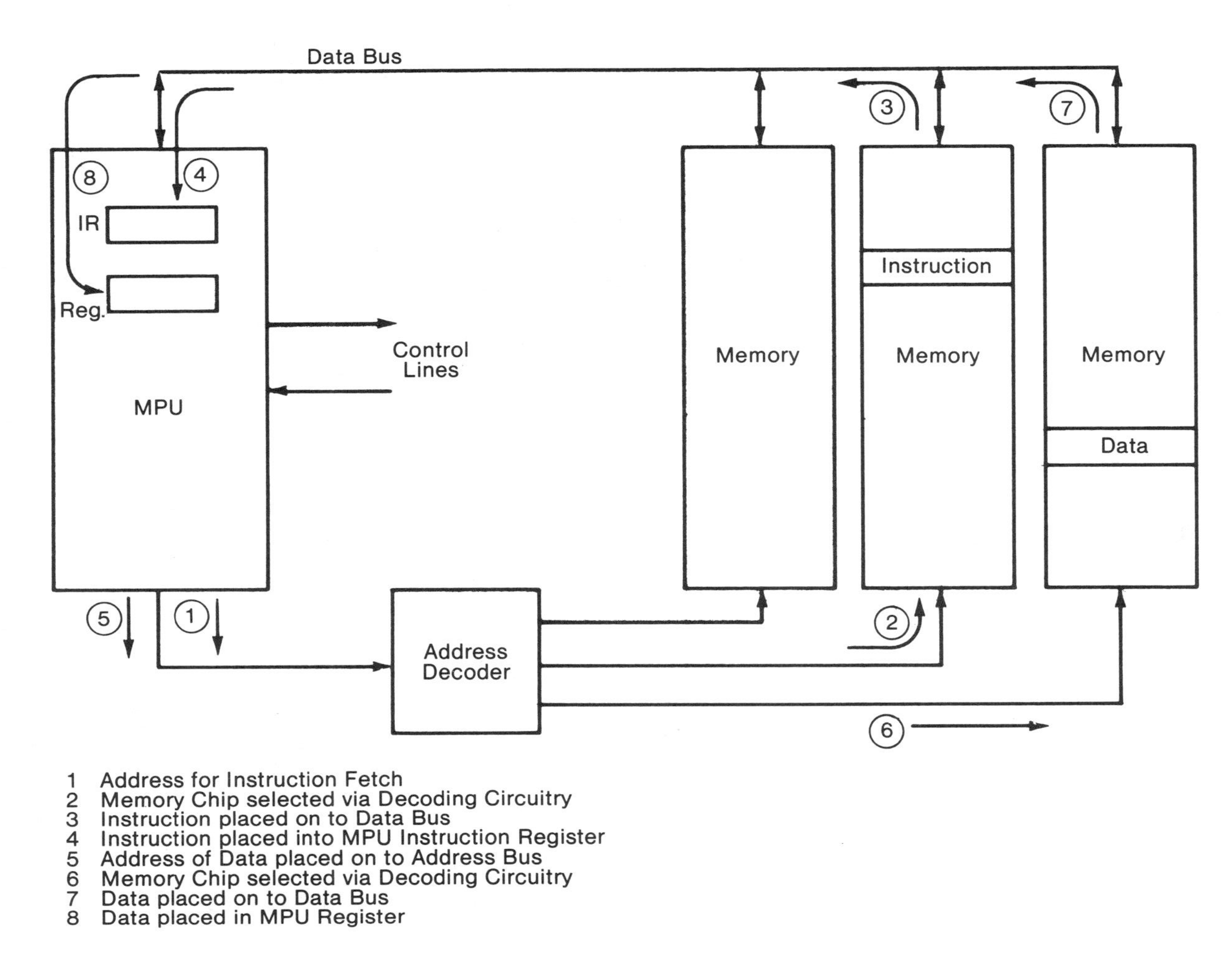

particular memory chip to be activated. The data in the addressed location is placed onto the data bus for acceptance by the CPU.

During the instruction fetch cycle portion of software processing, an address pointed to an instruction. During the execute cycle portion of the process, an address pointed to data. The process by which information was moved is the same in both instances, a flow from CPU to memory. All cycles in the CPU bear this resemblance. The CPU indicates where it expects to obtain or place information via an address. The data is then transferred over the data bus. In some cases data flows from the CPU to memory (a write operation) or from the CPU to/from input/output (I/O) devices. In all cases, the basic mechanism for the exchange of information is the same.

In some cases, the memory chips used in a computer system are not fast enough, in terms of access time, to keep up with the speed of the processor. This also applies to other peripheral chips used in microprocessor systems. This means that during the course of a memory transfer the slower memory chips would not have made their data available when the CPU was ready for it. Conceivably, the CPU would grab "garbage" from the data bus because the proper data just wasn't there yet. Since this is unacceptable, what can be done? An obvious solution is to use faster (and more expensive) memory chips. This can run up the cost of a system. Often, *wait states* are used instead. The slower memory chip (or other peripheral device) has the provision to signal the CPU that a wait state is necessary. When placed in the wait state, the CPU will maintain all signal levels but temporarily suspend further processing. This gives the slower component the time to complete its internal processes. After a sufficient time lapse, the slower chip removes the request for the wait state and normal CPU processing cycles continue. Each wait state lasts one clock period and will continue indefinitely if the signal requesting the wait state remains asserted. Wait states are also used with some peripheral chips to assure proper operation.

CLASSIFYING THE MICROPROCESSOR INSTRUCTION SET

The average microprocessor instruction set includes a large number of instructions, also known as *op codes* (operation codes), for a variety of purposes. Common to the majority of microprocessors are instructions designed to simply move data around the CPU and its associated memory. These instructions are often classified as belonging to a *data transfer group* of instructions and include the means to move, copy, exchange, load, or store data between registers and memory locations. Typically, these instructions comprise the majority of instructions in the instruction set.

- *Arithmetic instructions* include add, subtract, increment, and decrement commands in low-end processors. More advanced processors may include multiply, divide, and other complex mathematical algorithms. Some microprocessor families have mathematical coprocessors designed to run in tandem with the main processor. These high performance chips are beneficial for high speed, number crunching tasks.
- *Logical instructions* provide Boolean mathematical functions (AND, OR, NOT) and also include comparing, rotating, and shifting capabilities.
- *Branching instructions* can alter the normal flow of software execution and provide decision making capabilities. Jumps, calls, and returns are a few of the branching instructions available in most microprocessor instruction repertoires.
- An advanced processor may have a group of instructions for *string manipulation*. These instructions bear a resemblance to instructions in the data transfer group in that they also move data around the microprocessor system. The difference lies in the amount of data moved by a single instruction. A string of data is many bytes long and string manipulation instructions provide an easy mechanism for starting, moving, and stopping this type of data transfer.
- Finally, some instructions are dedicated to machine control, such as interrupt masking instructions, I/O instructions, and stack control instructions. These are generally referred to as *processor control instructions.*

In a typical microprocessor, a specific task can be performed by using one of several instructions. The wide range of instruction choices available to do seemingly similar procedures simply gives the programmer the versatility to create software applications that are fast, efficient, and elegant.

How the Microprocessor Processes Instructions

Instructions, when stored in memory, occupy a certain number of bytes; the number depends upon the particular instruction being used. For instance, an instruction that moves data from one internal CPU register to another will only be one byte long (Fig. I.4). When the CPU receives this instruction from memory and decodes it, the 8-bit pattern of ones and zeros uniquely defines the two registers involved in the transfer as well as the direction in which the data is to be transferred. No further information is needed to carry out this task once the decode is complete. Internal CPU timing takes over, finishes the job, and then requests the next instruction. If the data transfer involves a register and a memory location, then an address also needs to be specified. This address is provided to the CPU in several possible ways. The address could be stored in an internal register pair that the CPU merely references as part of the instruction execution. In this case the instruction itself will still be only one byte long since all necessary information for the source and destination of the data is internal to the processor. The instruction could provide the address along with the specific operation (Fig. I.5). In this instance the instruction would occupy three bytes of memory: one byte is the op code, which details that the operation to be carried out will involve moving data between a specific register and a memory location, and the two additional bytes contain the 16-bit address of the memory location involved.

All of the instructions in a typical microprocessor instruction set generally range from one to three bytes in length. In the two examples cited, a difference in execution speed is also evident. Each accessed byte in memory will entail a CPU memory reference cycle to set up the address of the location, establish the proper control signals, and access the data. This sequence of events takes several CPU clock cycles to complete. Therefore, multibyte instructions require more execution time than single byte instructions. The software application will determine if any of this is critical or not. If the application is not speed sensitive, then any approach to the software will most likely suffice. When speed is a factor, however, a few bytes saved here and there can make the difference between a slow, sluggish routine or a fast, responsive one.

Incidentally, the length of an instruction is related to the addressing mode used. Various addressing capabilities are designed into a microprocessor because of the versatility they bring to programming. The address mode used will provide the CPU or external device with the

Figure I.4 One-byte Instruction

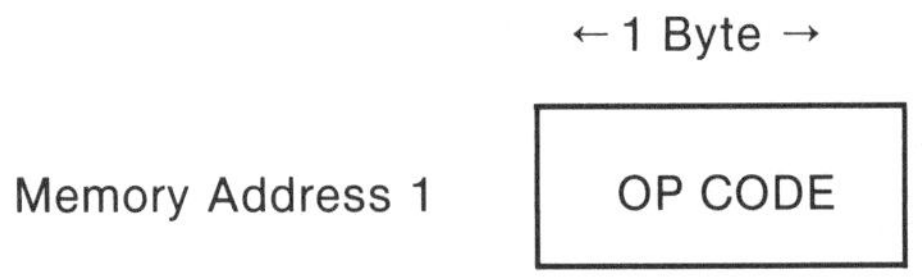

	← 1 Byte →
Memory Address 1	OP CODE

Figure I.5 Three-byte Instruction

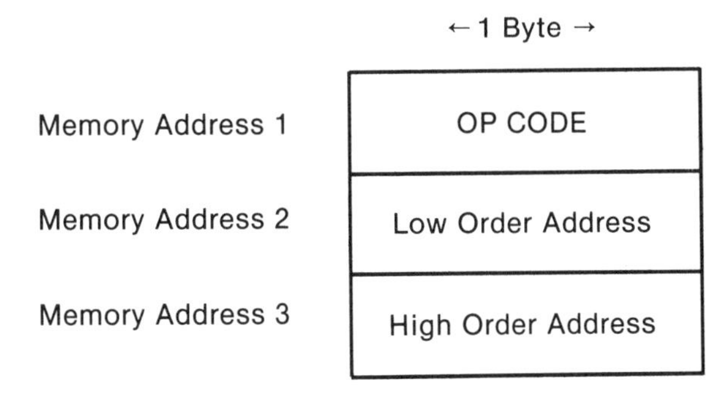

address of the data necessary for a particular operation. For example, *direct addressing* means that the address for a piece of data (often called the operand) is given directly with the instruction. The three-byte instruction example given earlier is an example of this. An *immediate addressing mode* instruction would have the data included with the instruction. In other words, the data is a part of the program. A constant value is often presented to the CPU with this addressing mode. Many other modes are designed into microprocessors, including indirect, register, absolute, and offset modes. The actual use of a particular mode comes about from the instruction chosen. The programmer needs to understand the modes and how effective addresses are calculated in order to correctly access data for processing.

Input/Output Processing

Information to be processed, or information that is the result of processing, is made available to the devices beyond the immediate confines of the CPU through I/O (Input/Output) channels. Devices referred to as input or output components run the gamut from printers, keyboards, and CRTs to sensors, valves, and lights. Many of these devices are not directly compatable with the electrical requirements of the CPU and, therefore, are attached to the system through an interface. The interface will convert voltage and current levels to appropriate digital levels as well as provide other functions such as data storage and signal gating. Data is transferred along the data bus in a manner similar to a memory data transfer. Since many I/O devices can be attached to the system (a function of the CPU used), they must be recognized by a unique address or *port number*. During the exchange of data information, the proper device address is placed on the address bus for this purpose. In a typical 8-bit microprocessor system, this address, or device code, is represented by an 8-bit field. This means that there can be 256 input devices as well as 256 output devices. Data that is destined for a particular output device is first loaded into a register (often the accumulator), and then the output command, along with the appropriate device code, is issued. During the execution of the command the CPU will place the register data on the data bus and the device code on the address bus. The selected interface circuitry then latches the data, making the output transfer complete. An input operation functions in a similar fashion.

Some processors also use a *memory-mapped* I/O technique. In this case, the address of an I/O device is represented as one of the addresses in the microprocessor's address space. Using this technique, more devices can be attached to the system since there are more available addresses. In addition, any software instruction that can move data to or from a memory location can be used. This gives the programmer greater versatility for efficient I/O operation. The disadvantage is that valuable memory space is traded off for increased I/O space. Further, a 16-bit (or greater) address must be decoded as part of the interface circuitry, which translates into additional hardware.

I/O operations involve the transfer of data between the CPU and external devices. The microprocessor peripheral chips available are designed to match a specific microprocessor family and to minimize the amount of necessary I/O interface hardware.

Interrupts

Interrupts save time. With this advantage, it is no wonder that interrupt capabilities are of primary interest when selecting a microprocessor. An interrupt structure will give processing time to the most important tasks in a computer system while still giving less demanding tasks the option to be serviced as needed.

Most processors have an internal interrupt flip-flop that will enable or disable the interrupt system. When enabled, the CPU is ever-vigilant for an interrupting signal from a device desiring service. With the occurrence of an interrupt signal, the CPU completes the currently executing instruction, disables further interrupts, and then responds to the specific interrupting device. Since many devices can cause an interrupt and since they all require a wide variety of software to handle their particular needs, some provision must be made to point the processor to the exact sequence of instructions for the interrupt at hand. When the interrupt first occurs, the processor could check with all the devices in the system until it finds the one that initiated the interrupt. This is referred to as a *polled interrupt.* This interrupt is rather simple and straightforward for smaller systems, but becomes cumbersome and slow as the number of devices increases. A more advanced technique involves the generation of a pointer or *vector* provided by the interrupting device, which will be accepted by the CPU and used to call the interrupt handling software. The procedure for doing this is very similar to a subroutine call. Many processors also have dedicated interrupt pins that, when asserted, will automatically call the proper software. These lines have priority assigned to them, which can be an advantage. Through preprogramming, active levels or edges can be assigned to these lines to simplify the associated hardware. Some lines, labeled *trap* or *non-maskable interrupt,* will have the highest priority over all others.

Interrupt controller chips are usually used to expand the interrupt structure in microprocessor systems. These chips are expandable, usually have eight interrupt request lines, can be configured through programming, and provide selective masking of interrupts.

INCREASING PERFORMANCE THROUGH DIRECT MEMORY ACCESS (DMA)

To illustrate the benefits of DMA, assume that a floppy disk drive is beginning to transfer data to the local memory area of a microprocessor-based computer system. The transfer could take place in the following fashion: The floppy disk controller places the data, one byte at a time, onto the data bus, where it is then captured by the CPU. The CPU temporarily stores this data in a register and then initiates an action that will place the data back onto the data bus for storage to memory. In order for this to occur, the CPU must, of course, provide an address, as well as the proper control signals. This series of events will be repeated for each byte transferred.

The procedure described will certainly accomplish its intention. However, in situations where data is to be transferred to or from a high speed device, it is not the best method available. That's where direct memory access comes in. In the situation described, all data has to move through the CPU on its way to memory. Other than providing the address and control signals, the CPU simply gets in the way of the data and slows it down on its way to memory. The DMA technique bypasses the CPU, provides its own memory addresses and control signals, and consequently, moves the data at a much greater rate of speed.

DMA is usually implemented through the use of a DMA controller chip, which supports these functions. A device needing or sending data does so through the DMA controller. The controller notifies the CPU of a DMA request which will ultimately place the CPU into a *hold state.* The CPU acknowledges that it is in the hold state with a signal typically called hold acknowledge (HACK). In the hold state, the CPU has electrically removed itself from the address and data buses by tristating its connection to these lines. This allows the DMA controller to access and control these buses. To complete the transfer to memory, the DMA controller provides addresses, data, and control signals. Using DMA, system throughput is dramatically improved.

SECTION II

Microprocessor Peripheral Integrated Circuits

II.1

Clock Generator and Controller

PART NUMBER	Z8581
FUNCTION	CLOCK GENERATOR AND CONTROLLER
MANUFACTURER	ZILOG
VOLTAGES	Vcc = +5
PWR. DISS.	750 mW
PACKAGE	18-pin DIP
TEMPERATURE	0°C → +70°C
FEATURES	TWO INDEPENDENT 20 MHz OSCILLATORS THREE CLOCK OUTPUT SOURCES STRETCHABLE CLOCKS - EXTERNALLY PROGRAMMABLE ON-CHIP RESET LOGIC
COMPATIBLE MICROPROCESSORS	Z80 Z8000
FUNCTIONAL DESCRIPTION	THIS CHIP PRODUCES THE TIMING SIGNALS REQUIRED IN A MICROPROCESSOR SYSTEM. ONE OSCILLATOR WILL PRODUCE TWO SYSTEM CLOCKS WHILE THE OTHER PRODUCES A GENERAL PURPOSE CLOCK OUTPUT. ONE OF THE SYSTEM CLOCK OUTPUTS MAY BE STRETCHED (ON THE HIGH OR LOW LEVEL) BY EXTERNAL CONTROLLING INPUTS.

Figure II.1.1 Z8581 Pinout

Reproduced by permission. 1986 Zilog, Inc. This material shall not be reproduced without the written consent of Zilog, Inc.

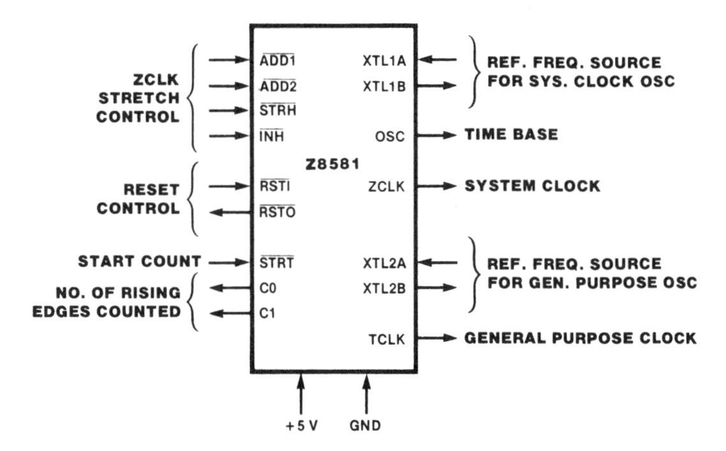

Figure II.1.2 Z8581 Block Diagram

Reproduced by permission. 1986 Zilog, Inc. This material shall not be reproduced without the written consent of Zilog, Inc.

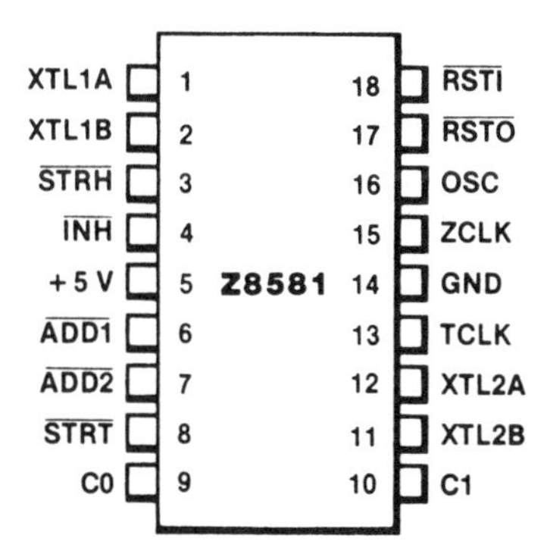

PIN NAME	PIN SYMBOL	PIN NUMBER	FUNCTION
SYSTEM CLOCK FREQUENCY SOURCE	XTAL1A XTAL1B	1 2	The crystal controlling the System Clock frequency is connected to these pins.
GENERAL PURPOSE CLOCK FREQUENCY SOURCE	XTAL2A XTAL2B	12 11	The crystal controlling the General Purpose Clock is connected to these pins.
TIME BASE CLOCK	OSC	16	TTL output with the same frequency as the System Clock Frequency Source.
SYSTEM CLOCK	ZCLK	15	This is the System Clock output signal. This pin drives the clock input of a microprocessor. The clock signal present at this pin can be modified by other control inputs described below.
GENERAL PURPOSE CLOCK	TCLK	13	This signal is one-half the frequency of the General Purpose Clock Frequency Source.
ADD DELAY 1,2	$\overline{ADD1}$ $\overline{ADD2}$	6 7	These two control lines add delay time (stretch the period) of the ZCLK output.
INHIBIT DELAY	$\overline{INH}$	4	$\overline{INH}$, when low, will inhibit the function of the $\overline{ADD1}$ and $\overline{ADD2}$ lines.
DELAY ZCLK	$\overline{STRH}$	3	When low, this line overrides the Add Delay and Inhibit Delay lines, and

PIN NAME	PIN SYMBOL	PIN NUMBER	FUNCTION
			will cause the ZCLK output to remain in its current state until $\overline{STRH}$ is inactivated.
START COUNT	$\overline{STRT}$	8	This input, when low, will reset an internal two-bit counter and then enable it for a counting sequence.
ZCLK COUNT 0,1	C0 C1	9 10	C0 and C1 are the outputs of the two-bit counter. The counter tracks the number of positive going edges that have occurred at ZCLK.
RESET IN	$\overline{RSTI}$	18	This is the input line for the RESET function. (Active low).
RESET OUT	$\overline{RSTO}$	17	Reset Out can be used as the microprocessor RESET input. This output is held low for a suitable number of clock periods for both power-on resets and $\overline{RSTI}$ conditions.
+5 V	+ 5 V	5	Supply voltage.
GROUND	GND	14	Supply ground.

The sequence of events initiated by a microprocessor is in itself derived from a continuously running master system clock. The master system clock is usually produced from a crystal-controlled oscillator which may or may not be a part of the microprocessor chip. For microprocessors that do not contain the necessary oscillator circuitry, a separate clock generator chip is used. Microprocessor systems built around processors that have the oscillator on board may also benefit from the enhancements supplied with a clock generator chip.

Z8581 OSCILLATORS

Three clock outputs are available on the Z8581. These clocks are produced from two, independent on-chip oscillators. The oscillators are controlled by external timing elements, such as crystals, or by an external frequency source. One of the oscillators is referred to as the System Clock Oscillator while the other is called the General Purpose Oscillator. The purpose of the System Clock Oscillator is to provide the timing signals necessary to drive a functioning microprocessor. Two clocks are actually available for this purpose. The Time Base Clock output (OSC) is a TTL compatible clock running at the same frequency as that set (see Fig. II.1.1.) by the crystal controlling the XTAL1A and XTAL1B inputs. Sufficient drive capability is available at this output to eliminate the requirement of external drivers and buffers. The other output, at one half the frequency of the System Clock Oscillator, is the ZCLK (System Clock) output. Timing pulses at this pin can be stretched (period extended) by control pins on the chip designated for this purpose. This provision will allow the system clock in a microprocessor system to be lengthened, under clock chip control, to accommodate slow memory, I/O, or peripheral chips. This method is an alternative solution to the addition of wait states in a microprocessor machine cycle and, conceivably, could provide superior performance. While a wait state will extend timing by a complete clock cycle at minimum, stretching a clock pulse can extend the pulse by as little as a single oscillator cycle. The shorter delay made possible by stretching a clock pulse can be an advantage in certain design situations.

A General Purpose Clock output (TCLK) runs at one-half the frequency of the General Purpose Clock Oscillator. Input pins XTAL2A and XTAL2B determine this frequency. This output is designed to provide the often necessary timing requirements that are not directly related to the system clock. For instance, a baud rate generator could easily be obtained using the TCLK output, eliminating the need for a dedicated chip to serve this purpose. Sufficient drive capability is available at all Z8581 outputs to eliminate the requirement for external drivers and buffers.

STRETCHING THE ZCLK

The majority of the pins on the Z8581 are dedicated to the control of the ZCLK output. These pins are grouped as ZCLK STRETCH CONTROL, START COUNT, and NO. OF RISING EDGES COUNTED as shown in Figure II.1.2. A fundamental way to stretch the System Clock is to hold input Delay ZCLK ($\overline{\text{STRH}}$) low. This will cause the ZCLK to remain in its current state until the STRH input is raised high. ZCLK can, therefore, remain in its current cycle, thus allowing slower devices extended time. Figure II.1.3 illustrates this point. The $\overline{\text{STRH}}$ control line overrides all other stretch control inputs.

Two other control inputs, Add Delays 1 and 2 ($\overline{\text{ADD1}}$, $\overline{\text{ADD2}}$) selectively delay the ZCLK output. These lines are effective only if enabled (high level) by the Inhibit Delay ($\overline{\text{INH}}$) line. Various binary combinations on the $\overline{\text{ADD1}}$ and $\overline{\text{ADD2}}$ lines will extend the current ZCLK half cycle by one, two, or three oscillator cycles. The levels on $\overline{\text{ADD1}}$ and $\overline{\text{ADD2}}$ are sampled prior to the rising edge of a System Clock Oscillator pulse. The delay specified by the control lines goes into effect at this rising edge. Figure II.1.4 illustrates how $\overline{\text{ADD1}}$ and $\overline{\text{ADD2}}$ can be used to add one and two additional oscillator periods to the ZCLK. Figure II.1.5 summarizes the effects of the stretch control lines.

A 2-bit counter, internal to the Z8581, monitors the status of oscillator cycles. The counter is reset and enabled by bringing the input Start Count ($\overline{\text{STRT}}$) low. The next four low-to-high transitions of ZCLK are then recorded by the counter. The current count is available at counter outputs C0 and C1. This provision is useful for programming a delay in ZCLK output. For example, assume that the third clock period in a microprocessor machine cycle is extended to allow completion of a memory reference utilizing slow RAM chips. This extension could be accomplished in the following fashion: Since most microprocessor chips have signal pins that indicate the beginning of a machine cycle

Figure II.1.3 ZCLK Stretched in the Low State

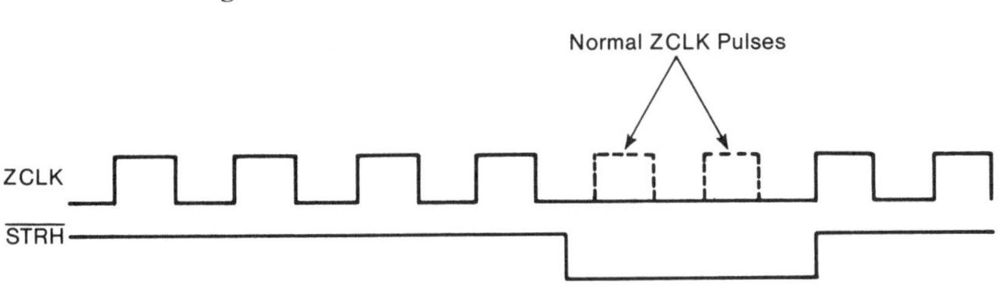

Figure II.1.4 Effect of $\overline{ADD1}$ and $\overline{ADD2}$ on the ZCLK Output

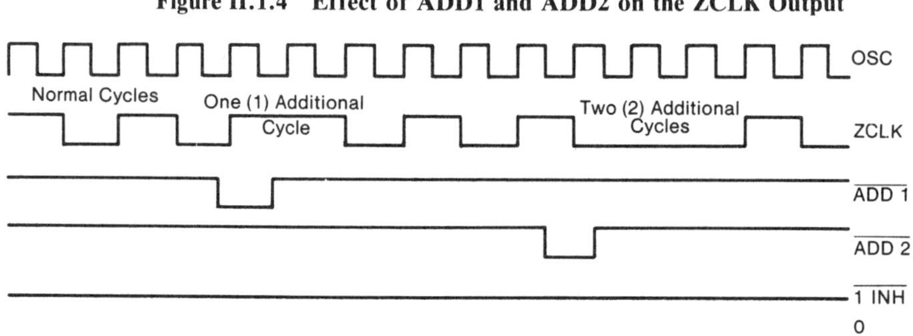

(i.e., M1 in the Z80), this signal could be connected to $\overline{STRT}$ to initialize the Z8581's 2-bit counter. The counter would then begin counting ZCLK's rising edges and, through a simple decoding of C0 and C1, machine cycle three could be detected. The decoded signal from C0 and C1 would be presented to $\overline{ADD1}$, causing the Z8581 to extend ZCLK by one oscillator clock period. Thus, machine cycle three is extended to allow the memory reference to complete.

RESETTING THE CLOCK GENERATOR

Two final pins on the Z8581 deal with the chip's reset capability. $\overline{RSTI}$ is the Reset input pin. When low, $\overline{RSTI}$ will cause output $\overline{RSTO}$ (Reset Out) to go low in sync with ZCLK's next positive going edge. $\overline{RSTO}$ can then be connected to the Reset input on the microprocessor. The pulse width of the $\overline{RSTO}$ line depends upon the circumstances that caused the Reset. For a normal reset, $\overline{RSTO}$ is brought low for 16 ZCLK cycles, the recommended time for Z80 and Z8000 microprocessors. During a power-on reset, the output is held low for 30 msec minimum. Internal power on reset and delay logic accomplish this.

Figure II.1.5 Stretch Control

Reproduced by permission. 1986 Zilog, Inc. This material shall not be reproduced without the written consent of Zilog, Inc.

$\overline{STRH}$	$\overline{INH}$	$\overline{ADD1}$	$\overline{ADD2}$	Periods Added
0	X	X	X	Unlimited
1	0	X	X	0
1	1	0	0	3
1	1	1	0	2
1	1	0	1	1
1	1	1	1	0

II.2

I/O Ports, Timer, and Static RAM

PART NUMBER	8155/8166 8155H/8156H 8155H-2/8156H-2
FUNCTION	256 × 8 STATIC RAM WITH I/O PORTS AND TIMER
MANUFACTURER	INTEL
VOLTAGES	Vcc = +5 Vss = GND
PWR. DISS.	725 mW
PACKAGE	40-pin DIP
TEMPERATURE	0°C → +70°C
FEATURES	256 × 8 STATIC RAM TWO 8-BIT PROGRAMMABLE I/O PORTS ONE 6-BIT PROGRAMMABLE I/O PORT 14-BIT BINARY COUNTER/TIMER
COMPATIBLE MICROPROCESSORS	8085AH 8085A 8088
FUNCTIONAL DESCRIPTION	THE 8155 IS A 256 × 8 STATIC RAM WITH PROGRAMMABLE I/O PORTS THAT MAY BE CONFIGURED AS INPUT, OUTPUT, OR CONTROL PORTS. THE 14-BIT COUNTER/TIMER CAN BE PROGRAMMED FOR SEVERAL MODES OF OPERATION. THE ACTIVE LEVEL OF CHIP SELECT IS THE ONLY DIFFERENCE BETWEEN THE 8155 AND 8156 PART NUMBERS. THE CHIP IS AVAILABLE IN VARIOUS PERFORMANCE GRADES.

Figure II.2.1 8155/8156 Pinout

Courtesy of Intel Corp.

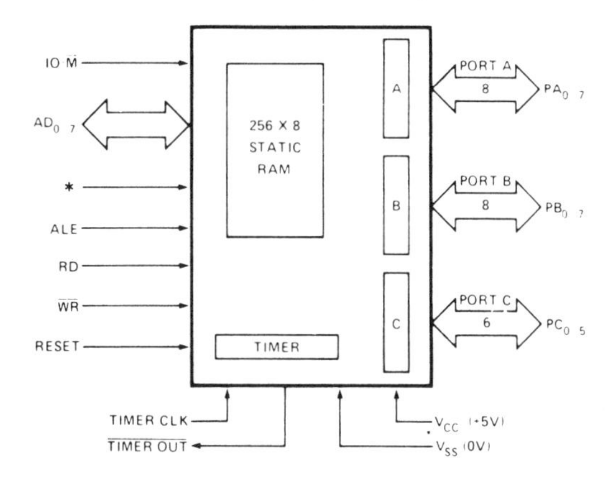

*8155H/8155H-2 = $\overline{CE}$, 8156H/8156H-2 = CE

Figure II.2.2 8155/8156 Block Diagram

Courtesy of Intel Corp.

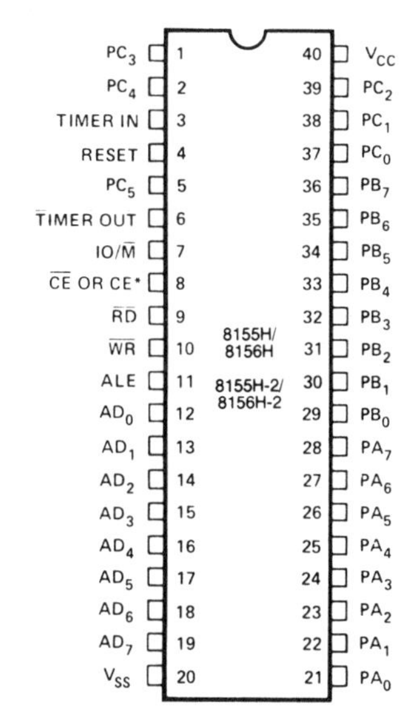

PIN NAME	PIN SYMBOL	PIN NUMBER	FUNCTION
RESET	RESET	4	A positive level signal on the Reset pin for two CPU clock cycles will reset the 8155/8156. This pulse typically comes from the CPU reset. Reset initializes the three I/O ports to input mode and stops the counter from counting.
ADDRESS/DATA 0 **ADDRESS/DATA 1** **ADDRESS/DATA 2** **ADDRESS/DATA 3** **ADDRESS/DATA 4** **ADDRESS/DATA 5** **ADDRESS/DATA 6** **ADDRESS/DATA 7**	AD0 AD1 AD2 AD3 AD4 AD5 AD6 AD7	12 13 14 15 16 17 18 19	These tri-state lines contain the address or data information from the CPU low order byte of the address/data bus. The 8155 latches the 8-bit address from these lines on the negative going edge of ALE. This address will select either RAM or I/O depending on the state of IO/$\overline{M}$.
CHIP ENABLE	$\overline{CE}$ or CE	8	When active, this signal enables chip operation. CE is active high for the 8156 and active low for the 8155.
READ CONTROL	$\overline{RD}$	9	Information from RAM or the various device registers will be read to the address/data bus when the chip is selected and $\overline{RD}$ is low.
WRITE CONTROL	$\overline{WR}$	10	Information for RAM or the various device registers is obtained from the address/data bus when the

PIN NAME	PIN SYMBOL	PIN NUMBER	FUNCTION
			chip is selected and $\overline{WR}$ is low.
ADDRESS LATCH ENABLE	ALE	11	This control line from the CPU will latch within the 8155/8156 the state of $\overline{CE}$, IO/$\overline{M}$, and the address from AD0–AD7. This occurs on the negative going edge of ALE.
I/O MEMORY	IO/$\overline{M}$	7	IO/$\overline{M}$ selects RAM if low or one of the various registers if high, when the chip is selected.
PORT A 0	PA0	21	These are the I/O pins for Port A (PA Register). Programming this port is accomplished through the Command Register.
PORT A 1	PA1	22	
PORT A 2	PA2	23	
PORT A 3	PA3	24	
PORT A 4	PA4	25	
PORT A 5	PA5	26	
PORT A 6	PA6	27	
PORT A 7	PA7	28	
PORT B 0	PB0	29	These are the I/O pins for Port B (PB Register). Programming this port is accomplished through the Command Register.
PORT B 1	PB1	30	
PORT B 2	PB2	31	
PORT B 3	PB3	32	
PORT B 4	PB4	33	
PORT B 5	PB5	34	
PORT B 6	PB6	35	
PORT B 7	PB7	36	
PORT C 0	PC0	37	These are the I/O pins for Port C (PC Register). Programming this port is accomplished through the Command Register.
PORT C 1	PC1	38	
PORT C 2	PC2	39	
PORT C 3	PC3	1	
PORT C 4	PC4	2	
PORT C 5	PC5	5	

(*continued*)

PIN NAME	PIN SYMBOL	PIN NUMBER	FUNCTION
TIMER IN	TIMER IN	3	Input pin for Timer/ Counter circuitry.
TIMER OUT	TIMER OUT	6	Output pin for Timer/ Counter circuitry. The output, either a pulse or square wave, is determined by programming the Count Length register.
SUPPLY VOLTAGE	Vcc	40	+5-V supply.
GROUND	Vss	20	Ground reference.

The 8155/8156 contains a small 256 × 8 static RAM memory but is probably more valuable as an I/O device. The combination of memory and I/O in the same chip package such as this, can significantly reduce chip count, particularly in smaller systems where memory and I/O requirements are at a minimum. This chip was designed to be compatible with several popular microprocessor chips, so that many of the control lines can be directly tied together, further reducing chip count. Since the chip functions both as a memory device and an I/O device, the IO/Memory Select (IO/$\overline{\text{M}}$) line found on the compatible microprocessors and the 8155/8156 (Figs. II.2.1 and II.2.2), is a key line toward differentiating whether a memory or I/O operation is taking place. With this line the 8155/8156 can be treated, via software, as a series of memory locations using standard memory reference instructions or as I/O ports using accumulator I/O instructions (IN and OUT). Both the ports and memory are functional at the same time.

USING THE 8155/8156 AS A MEMORY DEVICE

The RAM can be read or written from the driving microprocessor (CPU) with any memory reference instruction that specifies the address space assigned to the 8155/8156. The AD0 through AD7 pins are connected to the multiplexed lower eight bits of the CPU's address/data bus. These lines will contain address or data information at some point during the CPU's machine cycles. When address information is present on these lines, one of the 8155/8156 256 storage locations is being accessed, provided that the chip enable line ($\overline{CE}$) is active. This line is typically made active by decoding circuits attached to all or some of the high order eight bits of the CPU address bus. At some point in a CPU machine cycle, lines AD0 through AD7 are reserved for data transfers. Assuming that appropriate address information was obtained, the RAM will be read or written. The control lines that are used during this operation are $\overline{RD}$ or $\overline{WR}$; these indicate if data is going to or coming from the RAM; ALE, which latches the temporary address information from AD0 through AD7; IO/$\overline{M}$, which will be at the low level if a memory reference instruction is in progress.

USING THE 8155/8156 AS AN I/O DEVICE

The same lines controlling the RAM on the 8155/8156 are also used to control the I/O ports. The major difference in control is that line IO/$\overline{M}$ is high because the CPU will be processing an input or output instruction rather than a memory reference instruction. Under this condition, if the 8155/8156 is selected via the addressing circuitry, the I/O ports will be activated rather than the RAM.

There are three ports (registers) on the chip that are designated as PA, PB, and PC (Fig. II.2.1). Registers PA and PB are the 8-bit registers while PC is only 6 bits. Before the ports can be operated properly they must be programmed. This is done by modifying the contents of the Command Register (Fig. II.2.3) which controls the operation of PA, PB, and PC as well as the Timer. There is also a Status Register (Fig. II.2.4) that can be read by the CPU for information on the operation of the ports. The CPU can access these various registers through the address/data bus connection to AD0–AD7. To differentiate between registers, each register has been assigned a unique internal address so that they can be selectively modified (Fig. II.2.5). Most of the internal address is made up of don't-care bits since there are very few registers in the 8155/8156 compared to the 8 bits sent from the CPU that are used to access them.

The Command Register and the Status Register share the same address of XXXXX000. The Command Register can only be written to, and the Status Register can only be read from, so the $\overline{RD}$ and $\overline{WR}$ lines distinguish between the two. $\overline{RD}$ and $\overline{WR}$ are not active at the same time. The PA Register has address XXXXX001; PB is XXXXX010 and PC is XXXXX011. These addresses would be sent to the 8155/8156 from the CPU using IN or OUT instructions, so from

Figure II.2.3 Command Register Bit Assignment

Courtesy of Intel Corp.

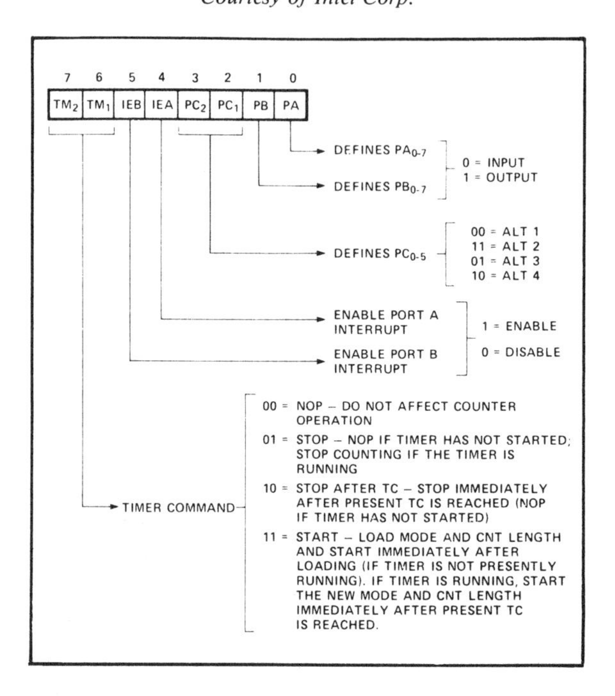

a programmer's perspective, these numbers constitute a device code or port address. The CPU places this address on the low order byte of the address bus as well as on the high order byte of the address bus. The lower order byte of the address bus from the CPU feeds the AD0 through AD7 pins on the 8155/8156. The high order byte of the CPU address bus typically feeds decoding circuitry that ultimately controls the chip enable ($\overline{CE}$) line. Therefore, in a circuit implementation using the 8155/8156, there will be a similarity between the address range assigned the chip and the port numbers of the chip.

For example, assume that the chip is assigned the address range 2000H through 20FFH (Fig. II.2.6). The two low order nibbles in this range account for the 256 bytes of RAM storage. If a memory reference instruction refers to one of the addresses within this range, the RAM will either be read or written. Assume further that the five most significant bits of the 20 portion of the address (high byte) feed the chip enable circuitry (ADDRESS/DATA bits A15–A11 = 00100, not an absolute decode). Either RAM or I/O ports may be selected within this range. The CPU via IO/$\overline{M}$ determines whether registers or memory is selected; the address sent to the 8155/8156 determines which particular

Figure II.2.4 Status Register Bit Assignment

Courtesy of Intel Corp.

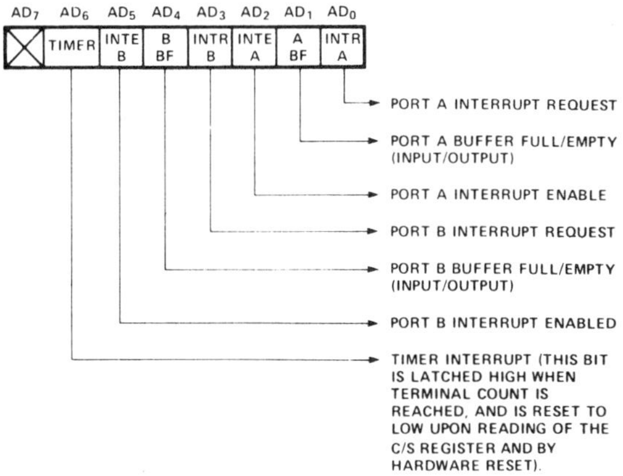

AD7	AD6	AD5	AD4	AD3	AD2	AD1	AD0
X	TIMER	INTE B	B BF	INTR B	INTE A	A BF	INTR A

- INTR A (AD0) → PORT A INTERRUPT REQUEST
- A BF (AD1) → PORT A BUFFER FULL/EMPTY (INPUT/OUTPUT)
- INTE A (AD2) → PORT A INTERRUPT ENABLE
- INTR B (AD3) → PORT B INTERRUPT REQUEST
- B BF (AD4) → PORT B BUFFER FULL/EMPTY (INPUT/OUTPUT)
- INTE B (AD5) → PORT B INTERRUPT ENABLED
- TIMER (AD6) → TIMER INTERRUPT (THIS BIT IS LATCHED HIGH WHEN TERMINAL COUNT IS REACHED, AND IS RESET TO LOW UPON READING OF THE C/S REGISTER AND BY HARDWARE RESET).

Figure II.2.5 I/O Port and Timer Addressing Scheme

Courtesy of Intel Corp.

I/O ADDRESS†								SELECTION
A7	A6	A5	A4	A3	A2	A1	A0	
X	X	X	X	X	0	0	0	Interval Command Status Register
X	X	X	X	X	0	0	1	General Purpose I O Port A
X	X	X	X	X	0	1	0	General Purpose I O Port B
X	X	X	X	X	0	1	1	Port C — General Purpose I O or Control
X	X	X	X	X	1	0	0	Low-Order 8 bits of Timer Count
X	X	X	X	X	1	0	1	High 6 bits of Timer Count and 2 bits of Timer Mode

X: Don't Care.

†: I/O Address must be qualified by CE = 1 (8156H) or $\overline{CE}$ = 0 (8155H) and IO/$\overline{M}$ = 1 in order to select the appropriate register.

Figure II.2.6 Addressing Example
Courtesy of Intel Corp.

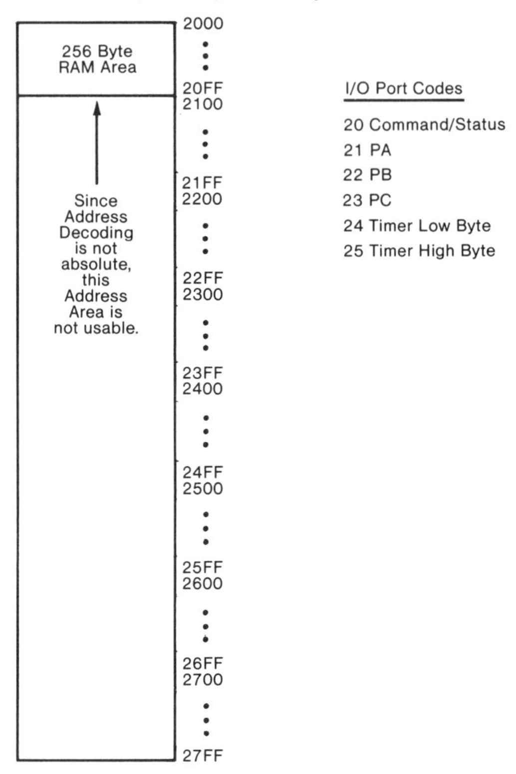

register or memory location is used. If the instruction being executed by the CPU is an IN or OUT instruction, the port number referenced by the instruction will be placed on the lower and upper portion of the address bus. Recall that the port number is only one byte long. The lower portion of the address bus feeds AD0 through AD7 while the upper portion, which has the same number as the lower portion for I/O instructions, partially feeds the chip enable circuitry as described above. The I/O ports would then have the following port numbers: Command/Status Register - 20H, PA - 21H, PB - 22H, PC - 23H. In addition, two registers to be described later for the timer would have the following port numbers: Timer high byte - 24H, Timer low byte - 25H.

Programming the registers for I/O operation first involves modifying the Command Register. Bits 0 and 1 in this register determine if PA or PB will be set up as input or output ports. Ones written to these two register bits will define registers PA and PB as output ports while zeros will configure the registers as input ports. Consequently pins PA0 - PA7 and PB0 - PB7, which are the I/O pins for these registers will respond in this fashion as well.

Bits 2 and 3 of the Command Register determine the mode of operation of register PC and its associated pins PC0 - PC5. A combination of 00 sent to these Command Register bits configures register PC as an input port. Combination 11 configures register PC as an output port. A combination of 01 configures register PC as a control port where pins PC1, PC2, and PC3 give status information on register PA. Pins PC3, PC4, and PC5 will become a 3-bit output port. Lastly, combination 10 to the Command Register configures PC0, PC1, and PC2 as a control register for port PA, and PC3, PC4, and PC5 as a control register for port PB. The port PC control register function is useful for handshaking purposes. The INTR line (PC0 or PC3) and BF (PC1 or PC4) information can also be obtained via the Status Register. Reading the Status Register while masking appropriate bits makes this information available to the CPU.

Bits 4 and 5 of the Command Register enable or disable interrupts for ports PA and PB. A one level in the register will enable the interrupt. Bits 6 and 7 of the Command Register control the operation of the Timer/Counter.

The Command Register can only be written. Some of the information in this register as well as information relating to other registers in the 8155/8156 can be obtained by reading the Status Register. Refer to Figures II.2.3 and II.2.4.

USING THE 8155/8156 COUNTER/TIMER

Two pins are used with the Timer circuitry. TIMER IN supplies a signal to the timer circuitry, while the output signal available at TIMER OUT depends on the programmed mode of operation of the timer.

The Timer has two internal addresses associated with it which are used to set up the desired count and mode of operation. The Command Register must also be set up for timer operation to proceed. Address XXXXX100 is the low order byte of the timer and address XXXXX101 is the high order byte. The low order byte contains bits T0 through T7 of the desired count length while the high order byte contains bits T8 through T13 (Fig. II.2.7). These 14 bits specify the length of the count. The remaining two bits in the high order byte, M1 and M2, determine the mode of timer operation. When Timer Mode bits M1 and M2 are

Figure II.2.7 Timer Format
Courtesy of Intel Corp.

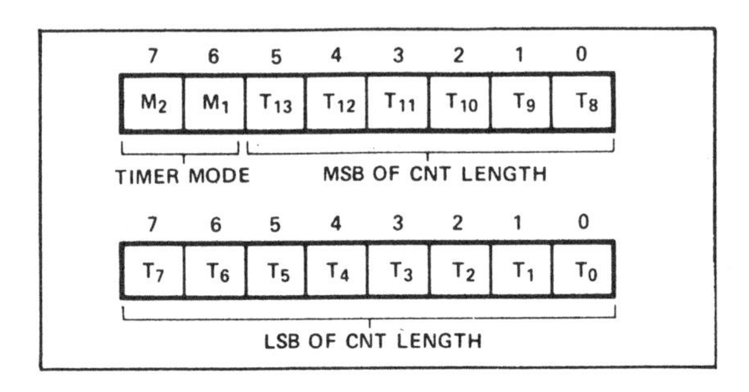

set to 00, the output at the TIMER OUT pin will be a single square wave (Fig. II.2.8) produced for a duration specified by the count length. This means that the square wave will start on the first count and end when the number of input counts at TIMER IN has reached the number specified by the count length. This is referred to as reaching the terminal count.

A continuous square wave can be generated at the TIMER OUT pin by setting the Timer Mode bits M1 and M2 to 01. The resulting square wave frequency depends on the input frequency at the TIMER IN input and the count length stored in the Timer's internal registers. For example, assume that the frequency of the signal applied to the TIMER IN pin is 3.125 Mhz and that the terminal count (count length) stored in the high and low byte addresses is 4138H. This number in binary is 00000100111000, which is equivalent to 312 decimal. This number divides down the frequency applied at TIMER IN. In this case the frequency of the square wave at TIMER OUT will be 3.125 MHz

Figure II.2.8 Timer Modes
Courtesy of Intel Corp.

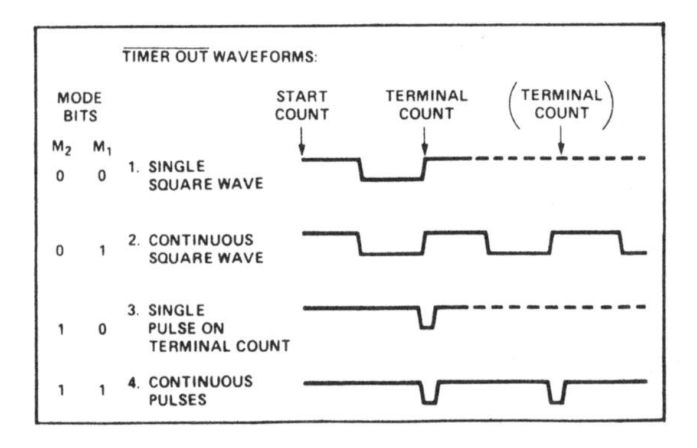

divided by 312 or 10 kHz. Any number from 2 to 16,383 decimal may be used to divide down the TIMER IN frequency.

Timer Mode bits M1 and M2, when set to 11, produce a continuous string of negative going pulses. The frequency of the pulse train is found in the same fashion as that for producing a continuous square wave. The pulse width will equal the period of the signal applied at TIMER IN. With a TIMER IN frequency of 3.125 MHz, the pulse width for the negative going pulse is measured as 0.32 μsec.

Timer Mode bits M1 and M2 set to 10 result in a single negative pulse when the terminal count is reached.

After the mode and count length have been set, the Timer needs to be started. A command loaded into the Command Register does this. There are four options available to control the Timer; they are selected by modifying bits 6 and 7 of the Command Register. Placing bits 6 and 7 to 00 is a NOP. Nothing happens to the Timer under this condition. This is a useful command for inhibiting Timer operation. Bits 6 and 7 set to 10 will stop the counter from running. Bits 6 and 7 set to 01 will stop the counter from running as soon as the terminal count is reached. Combination 11 will start the Timer by loading the count length and mode from the Count Length Register. If the counter is already running this load will take place as soon as the current terminal count is reached.

II.3

Dynamic RAM Controller

PART NUMBER	Am2964B Am2964C 2964B
FUNCTION	DYNAMIC MEMORY CONTROLLER
MANUFACTURERS	ADVANCED MICRO DEVICES SIGNETICS
VOLTAGES	Vcc = +5
PWR. DISS.	865 mW MAX
PACKAGE	40-pin DIP 44-pin LCC
TEMPERATURE	0°C → +70°C
FEATURES	CONTROLS BOTH 16K AND 64K DYNAMIC RAMS 8-BIT REFRESH COUNTER WITH CLEAR AND TERMINAL COUNT OUTPUT, 4 RAS OUTPUTS BURST, DISTRIBUTED, OR TRANSPARENT REFRESH MODES
COMPATIBLE MICROPROCESSORS	MOST PROCESSORS
FUNCTIONAL DESCRIPTION	A DYNAMIC RAM CONTROLLER FORMS THE HARDWARE INTERFACE BETWEEN A MICROPROCESSOR AND DYNAMIC RAM MEMORY CHIPS. THE CONTROLLER CHIP CONVERTS THE NORMAL ADDRESS BUS INFORMATION INTO A MULTIPLEXED FORM FOR THE DYNAMIC RAM AND ALSO PRODUCES THE NECESSARY CONTROL SIGNALS FOR PROPER OPERATION. IN ADDITION, DYNAMIC RAM CONTROLLERS PROVIDE THE SIGNALS NEEDED FOR RAM REFRESH.

Figure II.3.1 2964B Pinout

Courtesy of Signetics

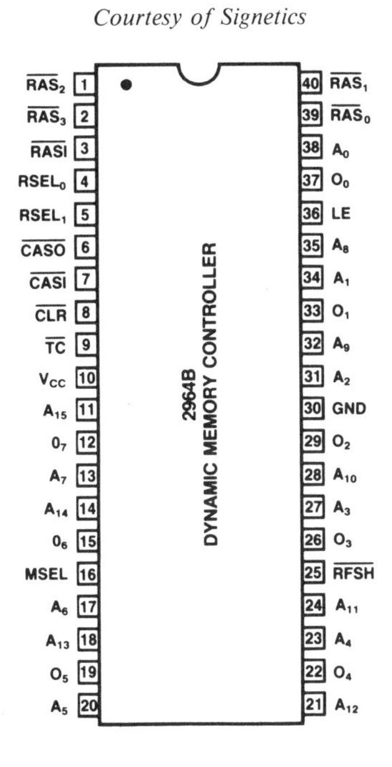

Figure II.3.2 Block Diagram of 2964B Dynamic Memory Controller (DMC)

Courtesy of Signetics

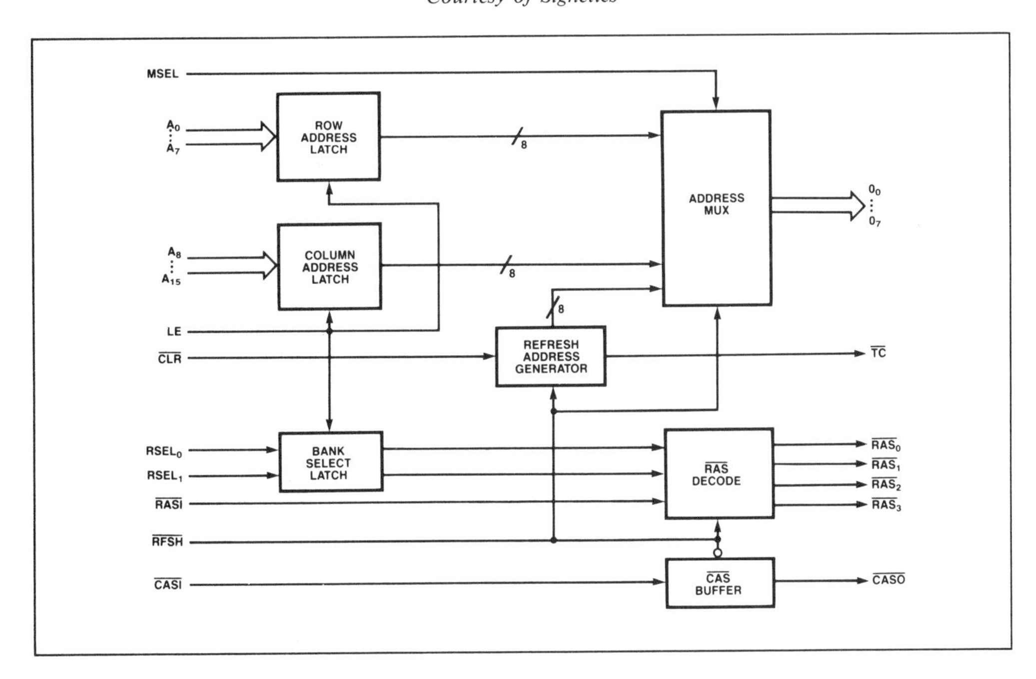

PIN NAME	PIN SYMBOL	PIN NUMBER	FUNCTION
ADDRESS BUS 0	A0	38	The microprocessor address bus lines are connected to the controller chip at these inputs. For dynamic RAM operation, the low order bits are used for row address inputs while the high order bits provide the column address. Bit A15 is designated as the most significant address bit for 64K chips; for 16K chips, A15 is tied to +12 V through a 1K resistor to alter the refresh counter $\overline{TC}$ output.
ADDRESS BUS 1	A1	34	
ADDRESS BUS 2	A2	31	
ADDRESS BUS 3	A3	27	
ADDRESS BUS 4	A4	23	
ADDRESS BUS 5	A5	20	
ADDRESS BUS 6	A6	17	
ADDRESS BUS 7	A7	13	
ADDRESS BUS 8	A8	35	
ADDRESS BUS 9	A9	32	
ADDRESS BUS 10	A10	28	
ADDRESS BUS 11	A11	24	
ADDRESS BUS 12	A12	21	
ADDRESS BUS 13	A13	18	
ADDRESS BUS 14	A14	14	
ADDRESS BUS 15	A15	11	
ADDRESS LATCH ENABLE	LE	36	A low level on this input causes the internal address latches and RAS Select latches to latch data presented to them. When high, the latches become transparent, letting data pass through them.
DRAM ADDRESS 0	O0	37	Normal memory addresses, both row and column, or a refresh address are presented to the Dynamic RAM (DRAM) on these output lines.
DRAM ADDRESS 1	O1	33	
DRAM ADDRESS 2	O2	29	
DRAM ADDRESS 3	O3	26	
DRAM ADDRESS 4	O4	22	
DRAM ADDRESS 5	O5	19	
DRAM ADDRESS 6	O6	15	
DRAM ADDRESS 7	O7	12	
MULTIPLEXER SELECT	MSEL	16	This control line, when high, places the address bus low order bits onto the dynamic RAM address output lines. When low,

PIN NAME	PIN SYMBOL	PIN NUMBER	FUNCTION
			the high order bits are routed to the output pins.
RAS SELECT 0 **RAS SELECT 1**	RSEL0 RSEL1	4 5	These inputs are latched under control of LE and are decoded to determine which memory bank is enabled through the activation of one of the $\overline{\text{RAS}}$ output lines.
ROW ADDRESS STROBE0 **ROW ADDRESS STROBE1** **ROW ADDRESS STROBE2** **ROW ADDRESS STROBE3**	$\overline{\text{RAS0}}$ $\overline{\text{RAS1}}$ $\overline{\text{RAS2}}$ $\overline{\text{RAS3}}$	39 40 1 2	The $\overline{\text{RAS}}$ outputs control selection of one of four memory banks. The lines are asserted low based upon the decode of inputs RSEL0 and RSEL1 when input $\overline{\text{RASI}}$ is low. All four lines are simultaneously active when both $\overline{\text{RASI}}$ and $\overline{\text{RFSH}}$ are low.
ROW ADDRESS STROBE INPUT	$\overline{\text{RASI}}$	3	Activation of the individual RAS outputs will occur when $\overline{\text{RASI}}$ is low. When both $\overline{\text{RASI}}$ and $\overline{\text{RFSH}}$ are low, all $\overline{\text{RAS}}$ outputs are active.
REFRESH CONTROL INPUT	$\overline{\text{RFSH}}$	25	In addition to activating all four $\overline{\text{RAS}}$ outputs, an active low $\overline{\text{RFSH}}$ places the refresh counter contents on the DRAM address lines and inhibits the $\overline{\text{CAS}}$ buffer.
REFRESH COUNTER CLEAR	$\overline{\text{CLR}}$	8	A low level on this input clears the refresh counter to zero. Due to inversion,

(continued)

PIN NAME	PIN SYMBOL	PIN NUMBER	FUNCTION
			the DRAM address outputs (O0–O7) will all read high upon asserting $\overline{\text{CLR}}$.
TERMINAL COUNT	$\overline{\text{TC}}$	9	This output reads low when the refresh counter has advanced to its highest count. The highest count will be 127 or 255, depending on the state of input A15.
COLUMN ADDRESS STROBE INPUT	$\overline{\text{CASI}}$	7	A low level on $\overline{\text{CASI}}$ will result in a low active level at $\overline{\text{CASO}}$, providing that a refresh cycle is not in progress.
COLUMN ADDRESS STROBE OUTPUT	$\overline{\text{CASO}}$	6	For normal memory operations, $\overline{\text{CASO}}$ is used to strobe the column address into the DRAM. A low level is active.
Vcc	Vcc	10	+5-V power supply.
GROUND	GND	30	Ground.

Dynamic RAM (DRAM) memory chips have a lower cost per bit, are physically smaller, use less power, and have greater density (bits per chip) as compared to static RAMs. Unfortunately, they also require a much higher degree of support circuitry to interface with a processor and to handle the peculiarities of memory refresh.

DRAMs are physically structured to save space by minimizing the number of input address pins so that, for instance, a 64K chip will have only eight address inputs rather than the 16 inputs normally required for 64K addresses. This necessitates two multiplexed addressing operations to provide the full 16-bit address. In this example, the address is broken into two segments of 8 bits each, called the Row Address and the Column Address. Individual control lines are needed to properly sequence the correct address portion to the DRAM at the proper time. DRAMs also have the interesting property that prevents them, on the average, from storing data for not much longer than 2 msec. This relatively short time period stems from the fact that the primary storage element within the DRAM is a capacitor that tends to leak off its charge. Thus the DRAM requires circuitry to provide a periodic refresh cycle which recharges the capacitors to their original levels. However, the packing density of the DRAM still makes it an attractive memory device. Fortunately, the speeds of current microprocessors allow plenty of time for useful processing activity to take place between the delays associated with refresh.

The dynamic RAM controller chip provides all these support functions, greatly minimizing the amount of hardware required for DRAM use.

2964 OVERVIEW

The 2964 Dynamic Memory Controller (DMC), as shown in Figures II.3.1 and II.3.2, can interface easily to 16K or 64K dynamic RAM chips. Address bit A15 is used as a normal address input for 64K chips. However, when using 16K DRAMs, A15 is tied to +12 V through a 1K resistor, altering the refresh counter $\overline{\text{TC}}$ output. The chip also has two inputs called RAS Select (RSEL0, RSEL1). These can function as high order address bits, allowing the DMC to control up to 256K of memory. RSEL0 and RSEL1 are decoded to produce four separate Row Address Strobe (RAS0-3) outputs. Each of these lines selects one of the four possible memory banks for use. If 64K chips are used, then four banks of these will yield a 256K system. Control lines Row Address Strobe input (RASI) and Refresh control input (RFSH), in conjunction with RSEL0 and RSEL1, determine when the Row Address Strobe outputs are activated. Figure II.3.3 summarizes the relationship between these lines.

The DMC chip has an 8-bit refresh counter whose purpose is to provide row addresses for the refresh cycle. The counter may be cleared at any time by bringing input $\overline{\text{CLR}}$ low. Normally the counter will advance one count each time either $\overline{\text{RASI}}$ or $\overline{\text{RFSH}}$ makes a transition from low to high. This allows each DRAM row to undergo a periodic refresh. A terminal count is reached and signaled on the Terminal Count ($\overline{\text{TC}}$) output pin whenever the counter has advanced through 256 states (max count of 255). When using 16K DRAMs, $\overline{\text{TC}}$ can be made to

Figure II.3.3 $\overline{\text{RAS}}$ Output Functions

Courtesy of Signetics

$\overline{\text{RFSH}}$	$\overline{\text{RASI}}$	RSEL_1	RSEL_0	$\overline{\text{RAS}_0}$	$\overline{\text{RAS}_1}$	$\overline{\text{RAS}_2}$	$\overline{\text{RAS}_3}$
L	H	X	X	H	H	H	H
L	L	X	X	L	L	L	L
H	H	X	X	H	H	H	H
H	L	L	L	L	H	H	H
H	L	L	H	H	L	H	H
H	L	H	L	H	H	L	H
H	L	H	H	H	H	H	L

indicate 128 states (max count of 127 and 255) through control of address input A15. The Refresh Counter will roll over and continue operation automatically as long as $\overline{\text{RASI}}$ or $\overline{\text{RASH}}$ pulses are provided.

How the DMC Responds to Memory Cycles

The DMC forms a hardware interface between a microprocessor and its memory. As such, the DMC must receive addresses from the CPU and modify them in a fashion appropriate for DRAM use. This means that a standard 16-bit address must be converted into an 8-bit row and an 8-bit column address. Since the DRAM will only have an 8-bit address input (assuming 64K DRAMs), the row and column addresses are not sent to the DRAM simultaneously. Control lines on the controller chip dictate the sequence of address flow. During a typical DRAM memory access, the DMC first places the row portion of the address on the 8-bit memory DRAM lines and then asserts a strobe line ($\overline{\text{RAS}}$) allowing the DRAM to internally latch this portion of the address. Shortly thereafter, the 8-bit column address is placed onto the DRAM address lines and latched by another strobe signal ($\overline{\text{CAS}}$). Read or write signals from the CPU are then produced, completing the memory cycle.

The DMC plays no part in regulating either read or write operations or the data lines. Rather, the DMC takes care of the address needs of the DRAM. Typically, the microprocessor has a control line that indicates when a valid address is on the address bus. This line could supply the DMC Address Latch Enable (LE) input with the signal to latch the 16-bit address from the CPU. An on chip multiplexer then places the low order part of the address, now referred to as the row address, on output address lines (O0-O7). Input line MSEL (Multiplexer Select) is raised high, allowing this to take place. A $\overline{\text{RAS}}$ (Row Address Strobe) line is then activated by the controller, which allows the DRAM to latch the row address. The 2964 has four $\overline{\text{RAS}}$ outputs, enabling the chip to drive up to four separate banks of memory. Choosing the desired $\overline{\text{RAS}}$ line is done by placing inputs RSEL0 and RSEL1 to the levels shown in Figure II.3.3. Also participating in this sequence is control input $\overline{\text{RASI}}$ (Row Address Strobe Input). This line, when low, actually allows the selected $\overline{\text{RAS}}$ output to become active.

All of the above activity simply presents the row address to the DRAM chip. The memory accessing operation cannot be completed until the column address reaches the DRAM. When the MSEL line is brought low, the internal multiplexer switches over to allow the column portion (high order part) of the original 16-bit address to pass through. The $\overline{\text{CASI}}$ (Column Address Strobe Input) is then brought low, in turn bringing controller output $\overline{\text{CASO}}$ (Column Address Strobe Output) low. This allows the DRAM to latch the Column address. There are two reasons for $\overline{\text{CASI}}$ to run through the DMC chip rather than run directly to the DRAM chips. First, in the event of a refresh operation, the $\overline{\text{CASO}}$ output is inhibited, even if $\overline{\text{CASI}}$ is active. Second, the on chip buffering of $\overline{\text{CASI}}$ minimizes the time delay differences between $\overline{\text{RAS}}$ and $\overline{\text{CASO}}$.

Once the DRAM has obtained all the necessary addressing information, the normal memory read or write operation progresses. Note that control inputs $\overline{\text{RASI}}$, $\overline{\text{CASI}}$, and MSEL may not come directly from the microprocessor. External sequencing circuits will most likely be required to produce these signals.

How the DMC Responds to Refresh

Refresh simply means restoring the DRAM capacitors to their original level of charge. This process is done by reading out the stored data and then rewriting it back to its original storage location. The internal DRAM circuitry and DRAM storage cell organization alleviate the hardware needs for this process as far as the controller chip is concerned. The reading and rewriting of data takes place within the DRAM. The cell organization (128 × 128 sections are typical) means that many cells are refreshed for each row address presented to the DRAM. This minimizes the number of refresh cycles needed to completely refresh the whole chip. Various refresh options, to be discussed later, are available to maximize overall system performance.

A refresh cycle is initiated when input $\overline{\text{RFSH}}$ is pulled low. At this point the refresh counter contents, rather than addressing information, pass through the chip multiplexer to the DRAM address inputs. Next, the $\overline{CASO}$ output is inhibited and all four $\overline{RAS}$ lines are activated as soon as $\overline{RASI}$ goes low. Placing the refresh counter output (inverted output, to be exact) at the DRAM address input allows one full row of cells to be refreshed. All four $\overline{RAS}$ outputs are asserted at the same time, allowing corresponding rows in all four memory banks to refresh on the same cycle. Figure II.3.4 illustrates the refresh timing waveforms.

At the end of the refresh cycle, the refresh counter advances one count on the positive going edge of either $\overline{RFSH}$ or $\overline{RASI}$. The refresh counter is then ready to provide the refresh address for the next refresh cycle. A Terminal Count ($\overline{TC}$) signifies that the counter has progressed through 256 (or 128) counts, meaning that all rows have undergone a refresh. Terminal Count is an optional DRAM Memory Controller output. The counter will return to a count of zero when the current cycle (the one that produced a $\overline{TC}$) is complete or upon the receipt of a $\overline{CLR}$ pulse.

REFRESH MODES

Several refresh modes provide flexibility for the memory system designer. Some external logic is required for full implementation of these methods.

Using the Distributed refresh method, all rows will be refreshed within the 2 msec time requirement but not with any particular timing relationship to normal memory cycles. Basically, this means that refresh cycles are executed periodically by momentarily suspending normal memory read and write cycles. One must insure that all DRAM rows are refreshed within the proper time span. The refresh timing is the same as that described above.

Transparent refresh also follows the described refresh cycle timing sequence. The difference between this method and Distributed refresh is that a Transparent refresh cycle occurs for every memory reference operation. Since many microprocessors spend some time during a memory reference internally determining the next course of action (i.e., fetch next op code, output data), the address bus becomes idle for a short period of time. If the controller can detect this idle time, then a refresh cycle can be inserted without any disruption in normal processing. The external circuitry required to support this method must examine processor status and tell the controller chip when to begin a refresh. Although this procedure can bury the time requirements of

Figure II.3.4 Refresh Timing Waveforms

Courtesy of Signetics

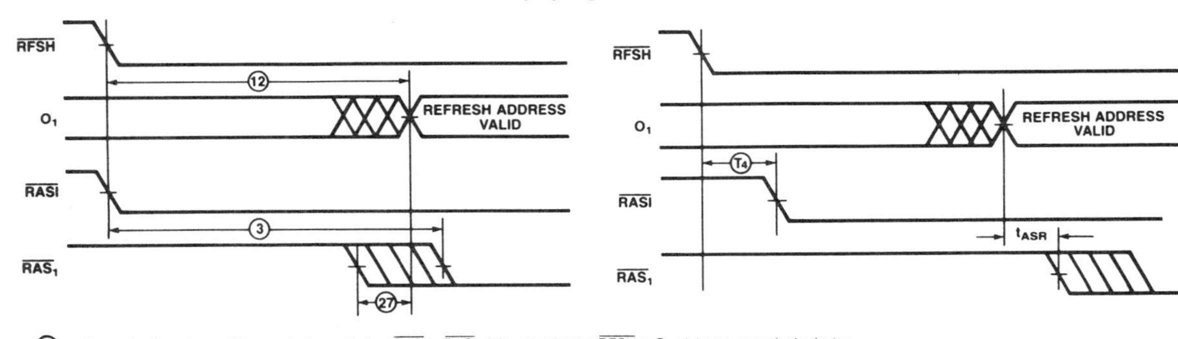

(27) = Guaranteed maximum difference between fastest $\overline{RASI}$ to $\overline{RAS_1}$ delay and slowest $\overline{RFSH}$ to O_1 delay on any single device.

a. Test Waveforms **b. Desired System Timing**

refresh within normal system timing, a problem can arise if the processor suspends normal activity. Situations such as a system halt or extensive DMA activity could hold up a refresh cycle to the point that data is lost. Also, some newer processors make use of the idle time for instruction prefetch and cannot use this refresh method.

A final method is Burst refresh. Burst refresh means that all rows are refreshed in sequence. To implement this with a 2964, $\overline{\text{RFSH}}$ is brought low for a period of time (many cycles). $\overline{\text{RASI}}$ is then pulsed 128 or 256 times to advance the refresh counter through all states and to activate the $\overline{\text{RAS}}$ outputs. All DRAM cells are refreshed sequentially with this method. A Burst refresh needs to take place approximately every 2 msec, and, when it occurs, stops all normal bus activity until the total refresh is complete.

II.4

Programmable Communication Interface

PART NUMBER	Am9551 8251 8251A
FUNCTION	PROGRAMMABLE COMMUNICATION INTERFACE
MANUFACTURERS	AMD INTEL NATIONAL NEC SMC WESTERN DIGITAL
VOLTAGES	Vcc = +5
PWR. DISS.	1 W max
PACKAGE	28-pin DIP
TEMPERATURE	0°C → +70°C
FEATURES	PROGRAMMABLE ASYNCHRONOUS AND SYNCHRONOUS OPERATION MAX ASYNCHRONOUS BAUD RATE = 19.2 kHz MAX SYNCHRONOUS RATE = 64 kHz FULL DUPLEX DOUBLE-BUFFERED RECEIVER AND TRANSMITTER PARITY, OVERRUN, AND FRAMING ERROR DETECTION
COMPATIBLE MICROPROCESSORS	MOST 8- AND 16-BIT PROCESSORS
FUNCTIONAL DESCRIPTION	THIS CHIP IS DESIGNED TO HANDLE DATA COMMUNICATIONS BETWEEN A MICROPROCESSOR AND A PERIPHERAL DEVICE. THE CHIP MAY BE PROGRAMMED BY A MICROPROCESSOR TO HANDLE BOTH SYNCHRONOUS AND ASYNCHRONOUS DATA TRANSMISSION FORMATS. THE TRANSMITTER AND RECEIVER CLOCKS ARE INDEPENDENT OF ONE ANOTHER. IN ASYNCHRONOUS MODE, THE BAUD RATE CAN BE VARIED BY PROGRAMMING.

Figure II.4.1 8251A Pinout

Courtesy of Intel Corp.

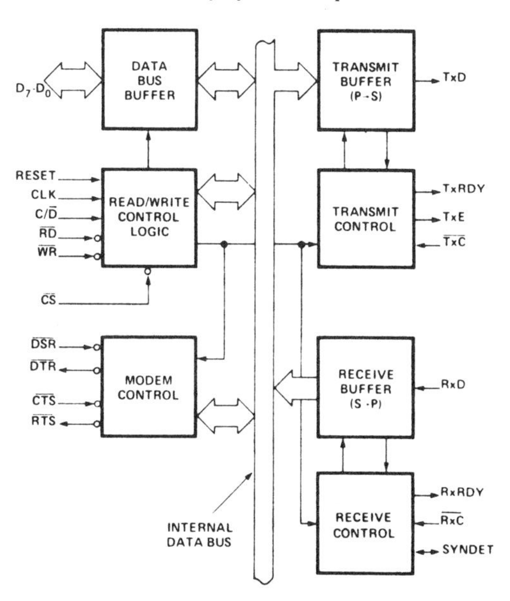

Figure II.4.2 8251A Block Diagram

Courtesy of Intel Corp.

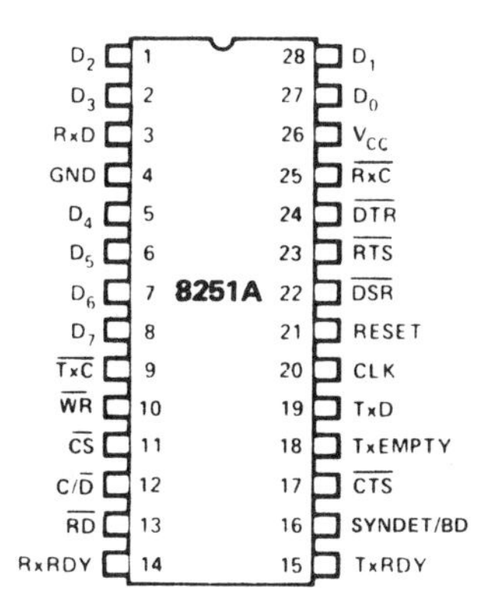

PIN NAME	PIN SYMBOL	PIN NUMBER	FUNCTION
RESET	RESET	21	A one level on this pin for at least six clock periods forces the chip into an idle state where no activity can take place. A set of new control words must then be issued by the CPU to reactivate the chip. The 8251 may also be reset by a software command.
CLOCK	CLK	20	This clock input controls only the 8251 internal timing. Data transfer timing is controlled via separate pins (9 and 25). The clock frequency is related to transmit and receive operation in that the CLK input must be at least 30 times the transmit and receive clock frequencies in 1 × mode and 4.5 times as great in all other modes.
DATA BUS 0 DATA BUS 1 DATA BUS 2 DATA BUS 3 DATA BUS 4 DATA BUS 5 DATA BUS 6 DATA BUS 7	D0 D1 D2 D3 D4 D5 D6 D7	27 28 1 2 5 6 7 8	The data bus is a tri-state, bidirectional bus connecting the 8251 to the controlling microprocessor. All data, control words, command words, and status information are transferred on these lines.
READ WRITE	$\overline{\text{RD}}$ $\overline{\text{WR}}$	13 10	When low (only one line can be low at a time) these two control lines from the CPU instruct the 8251 to receive information from or send information to the

PIN NAME	PIN SYMBOL	PIN NUMBER	FUNCTION
			CPU. The exchange of this information (data, commands, status) takes place on the data bus.
CHIP SELECT	$\overline{CS}$	11	The chip is selected by a low level on this line. When $\overline{CS}$ is high $\overline{RD}$ and $\overline{WR}$ have no effect. The data bus is tri-stated when $\overline{CS}$ is high.
CONTROL/DATA	C/$\overline{D}$	12	This line, in addition to the information on the $\overline{RD}$ and $\overline{WR}$ line, informs the 8251 that the byte of information on the data bus is either a control status character or data. C/$\overline{D}$ = 0 signifies data, while C/$\overline{D}$ = 1 signifies control/status.
TRANSMITTER CLOCK	$\overline{TxC}$	9	This clock controls the rate at which data is transmitted out of the USART at the transmit data output pin (TxD). Data is available at this pin on the falling edge of the $\overline{TxC}$ clock. The transmit clock frequency depends upon the mode of operation. In synchronous mode the baud rate is equal to the transmit clock rate. In asynchronous mode, the baud rate is a fraction of the transmit clock frequency

(*continued*)

PIN NAME	PIN SYMBOL	PIN NUMBER	FUNCTION
			(1, 1/16, 1/64 - set by the mode instruction).
RECEIVER CLOCK	$\overline{\text{RxC}}$	25	This clock controls the rate at which data is received by the USART at the receiver input pin (RxD). Data is accepted by the USART on the rising edge of the receiver clock. Baud rates are determined in the same manner as those for the transmitter clock.
TRANSMITTER DATA	TxD	19	TxD is the serial output line of the USART. Data is sent out on this line under control of the transmitter clock ($\overline{\text{TxC}}$) and other control lines. When data is not being transmitted, this output is typically in the high state (mark).
RECEIVER DATA	RxD	3	USART serial input data is received on this line under control of the receiver clock and other control lines.
TRANSMITTER READY	TxRDY	15	When high, TxRDY informs the host microprocessor that the transmitter section of the USART is ready to accept a data character. There are two ways to do this. The line TxRDY can be used to interrupt the microprocessor or, via a status read opera-

PIN NAME	PIN SYMBOL	PIN NUMBER	FUNCTION
			tion, the state of TxRDY can be sensed. TxRDY is reset on the negative-going edge of $\overline{WR}$.
TRANSMITTER EMPTY	Tx-EMPTY	18	TxEMPTY is high when the 8251 transmitter has no characters to send, when SYNC characters are being transmitted, or when the transmitter is disabled. When the transmitter is enabled and a data character is received from the microprocessor, then TxEMPTY is reset.
RECEIVER READY	RxRDY	14	When high, RxRDY indicates to the microprocessor that an assembled character is ready. The microprocessor can be informed that the character is ready by using RxRDY as an interrupt or by sensing the state of RxRDY via a Status read operation. When the microprocessor reads the character from the receiver, RxRDY is automatically reset.
DATA SET READY	$\overline{DSR}$	22	The $\overline{DSR}$ line is generally connected to a modem and is used to test modem conditions. The host microprocessor can sense the condition of this line

(*continued*)

PIN NAME	PIN SYMBOL	PIN NUMBER	FUNCTION
			through a Status read operation.
DATA TERMINAL READY	$\overline{DTR}$	24	This line is used to drive the Data Terminal Ready line on a modem. This output is controlled by a Command Instruction.
REQUEST TO SEND	$\overline{RTS}$	23	This output is also controlled by a Command Instruction and is typically used to drive a modem Request To Send line.
CLEAR TO SEND	$\overline{CTS}$	17	A low on this line allows the transmitter to send data if the transmitter is enabled. All data in the transmitter is sent in the event that the transmitter is disabled (by $\overline{CTS}$ or a Command) in the middle of a transmit operation.
SYNC/BREAK DETECT	SYNDET/ BRKDET	16	Through a Control Word, this pin can be configured as an input or output. As an output, SYNDET goes high to indicate that a SYNC character (synchronous operation only) has been detected. The line is automatically reset with a Status Read operation. As an output in asynchronous operation, this line goes high when an all-zero word of specified length is received.

PIN NAME	PIN SYMBOL	PIN NUMBER	FUNCTION
			BRKDET will return low on a Reset or when the Receiver Data line returns to a one level. The line can be programmed as an input in synchronous mode to allow for an external SYNC mode of operation.
Vcc	Vcc	26	+5 V
GROUND	GND	4	Ground.

Serial data communication can be accomplished using both synchronous and asynchronous techniques. The 8251 Programmable Communication Interface provides the capabilities to operate in these modes. The programmable nature of the chip allows for variable baud rates, character size, and stop bits, as well as error detection. The chip has both transmit and receive sections that can operate concurrently. Figure II.4.1 illustrates the internal structure of the 8251. Figure II.4.2 details the chip pinout.

UNDERSTANDING 8251 BASIC OPERATION

Figure II.4.3 illustrates the basic 8251 interface formed between the host microprocessor and a peripheral device. Between the microprocessor and the 8251 a number of lines are necessary to provide programming control and a path for the exchange of data. The RESET line will idle the 8251 in preparation for a Mode Instruction. The Mode Instruction will initialize the chip for synchronous/asynchronous operation, baud rate, number of stop bits, parity, and character length. The chip can also be reset via a software command. Assuming that the chip has been reset, the first write operation to the chip will be interpreted by the chip as the Mode Instruction. The Mode Instruction, as well as all Command Instructions, Status, and data, are exchanged between the 8251 and the microprocessor on the 8-bit data bus (D0-D7). The control and meaning of the information on the data bus are handled by several lines. $\overline{WR}$ and $\overline{RD}$ are the write and read control lines

Figure II.4.3 Basic USART Interface

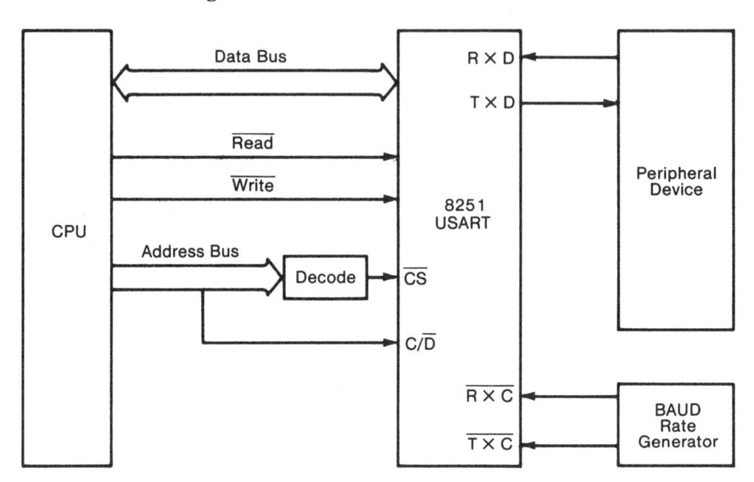

emanating from the microprocessor. These lines establish the direction of information flow between the two chips. Only one line can be active at a time and must be conditioned by Chip Select ($\overline{CS}$). To distinguish between Mode/Command/Status and data information, the Control/Data ($C/\overline{D}$) line is used. Typically, this line is tied to an address bit of the controlling microprocessor. When one address is specified by the microprocessor's controlling program, a data exchange takes place; another address is used for the exchange of control information.

The CLK line is used for internal timing in the 8251. This clock does not control the baud rate. However, the clock frequency must be greater than thirty times the Transmitter/Receiver rate.

PROGRAMMING THE USART

Immediately following a reset, the controlling microprocessor must load several control words into the USART. A simple write operation (with $C/\overline{D} = 1$) will accomplish the load. However, it is the sequence in which the control words arrive at the USART which determines their meaning. The first control word written to the USART is the Mode Instruction. The format for this control word is shown in Figure II.4.4. This instruction distinguishes between synchronous and asynchronous operation by selection of the Baud Rate Factor (bits D0, D1). Three bit combinations are labeled 1×, 16×, and 64×. One of these three combinations configures the USART for asynchronous operation. The

Figure II.4.4 Mode Instruction

Courtsy of Intel Corp.

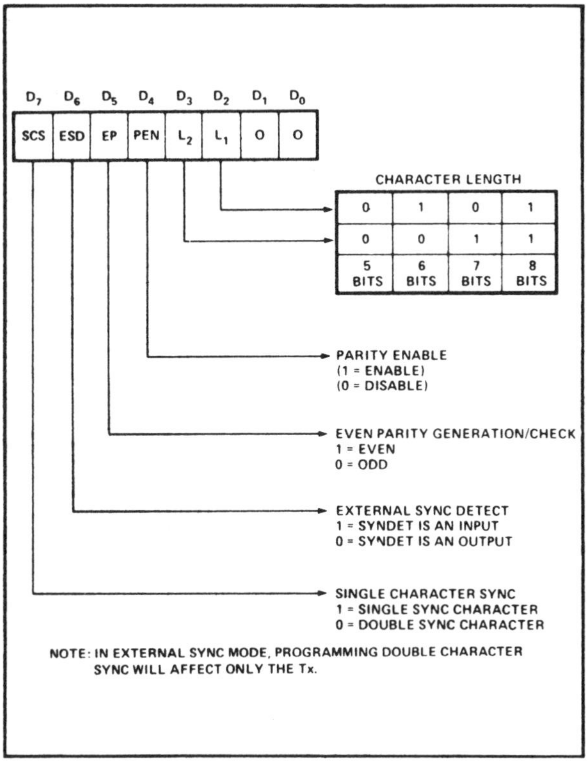

D_7	D_6	D_5	D_4	D_3	D_2	D_1	D_0
SCS	ESD	EP	PEN	L_2	L_1	0	0

CHARACTER LENGTH

L_1	0	1	0	1
L_2	0	0	1	1
	5 BITS	6 BITS	7 BITS	8 BITS

PARITY ENABLE
(1 = ENABLE)
(0 = DISABLE)

EVEN PARITY GENERATION/CHECK
1 = EVEN
0 = ODD

EXTERNAL SYNC DETECT
1 = SYNDET IS AN INPUT
0 = SYNDET IS AN OUTPUT

SINGLE CHARACTER SYNC
1 = SINGLE SYNC CHARACTER
0 = DOUBLE SYNC CHARACTER

NOTE: IN EXTERNAL SYNC MODE, PROGRAMMING DOUBLE CHARACTER SYNC WILL AFFECT ONLY THE Tx.

baud rate then becomes a multiple of the Transmit/Receive clock rate. For example, if the Transmit clock is equal to 600 Hz and 16 × is chosen in the Mode Instruction, then the resulting transmitter baud rate is 600 × 16 = 9600. If Sync Mode is selected, the USART is configured for synchronous operation with the baud rate equaling the Transmit/Receive clock rate. The remaining bits in the Mode Instruction are used to set up the characteristics of the transmitted/received data used in a serial data communications system.

The 8251 is designed to accept as the Mode Instruction the first write operation after a reset. Command Instructions are next written by the microprocessor to control the data transfer operations of the 8251. A Command Instruction always follows a Mode Instruction, although Command Instructions may be issued at any time. Figure II.4.5 details the Command Instruction Format. Command Instruction details are discussed in the following sections.

Figure II.4.5 Command Instruction

Courtesy of Intel Corp.

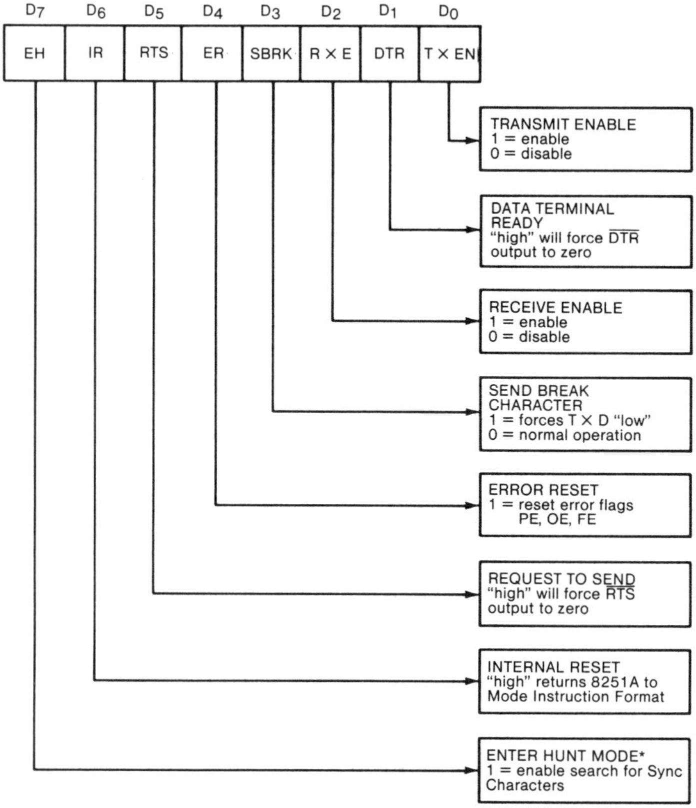

Note: Error Reset must be performed whenever R × Enable and Enter Hunt are programmed.

*(HAS NO EFFECT IN ASYNC MODE)

Using the USART for Asynchronous Operation

- *Data Transmission.* Data to be transmitted by the USART in serial form must first be loaded into the USART from the controlling microprocessor. The Command Instruction previously sent must have the Transmitter Enable bit (TxEN) set to one to enable the Transmitter for operation. In an asynchronous mode of operation, the data is loaded following the Command Instruction using a write operation. This time the write operation is done with the $C/\overline{D}$ bit equal to zero, indicating to the USART that the information sent from the microprocessor is character data. This data is sent to the USART in parallel on the data bus. The USART will automatically add a low level start bit to the data as well as the programmed number of stop bits and parity. This data is moved through the internal double-buffered data path, converted from parallel to serial form (start bit first), and transmitted out on the Transmitter Data output pin (TxD). The rate of transmission is 1, 1/16, or 1/64 of the transmitter clock present at pin $\overline{\text{TxC}}$, giving each bit of the data a bit time corresponding to a fraction of the Transmitter Clock. Data is shifted out on the negative going edge of the $\overline{\text{TxC}}$ clock. When data is no longer available for transmission, the transmitter data out pin will be placed into the high state.

- *Data Reception.* The USART receives serial data on the Receiver Data input pin (RxD). This line is controlled by the device sending data to the USART and is held in the high state until data is ready to be sent. The USART receiver will realize that data is available once a negative going edge is detected on the data input pin. The line going low is actually the start bit for the character of data that the USART is going to receive. To ensure that the negative-going edge is in fact the beginning of a valid start bit and not a noise glitch, the 8251 will recheck the start bit halfway through its bit time. If the line is still low, then the level is interpreted as a start bit and the reception of data begins. Data is received on the positive going edge of the receiver clock ($\overline{\text{RxC}}$). Parity, framing, and overrun errors are checked at this time. These errors do not inhibit 8251 operation. The microprocessor can detect these errors by sampling the Status Register (described later) on a periodic basis.

  When the USART has received a character of data the microprocessor must accept it before the next character arrives. The microprocessor can be notified of the presence of data via an interrupt. When data is available, the Receiver Ready (RxRDY) pin is brought to a high state by the USART, providing the means for an interrupt. Alternatively, the microprocessor can poll the USART by reading the level of bit D1 in the Status Register. This bit contains the same information as the RxRDY line.

Using the USART for Synchronous Operation

The 8251 is programmed for synchronous operation by setting the appropriate bits in the Mode Instruction (recall that this instruction always follows a reset). Immediately after this instruction the USART will accept one or two SYNC character bytes. Following the SYNC bytes, a Command Instruction sent to the USART will initiate synchronous operation. Prior to the actual transmission or reception of data, all this program information is loaded into the USART from the controlling microprocessor.

- *Data Transmission.* The Transmitter output pin (TxD) remains in the high state until the controlling microprocessor sends over the first SYNC character. The device that will be receiving data notifies the 8251 that it is ready for data by bringing Clear To Send ($\overline{\text{CTS}}$) low. Data is then serially shifted out of the USART at the same rate as the Transmitter Clock ($\overline{\text{TxC}}$). The SYNC characters are sent first with the data characters following. Unlike asynchronous transmission, several characters are sent at once. In fact, synchronous data transmission protocol requires a steady stream of data in order to maintain synchronization between the transmitter and receiver. Therefore, the microprocessor must keep a supply of data characters available to the USART. If data is unavailable, the USART will automatically insert the preprogrammed SYNC characters into the data stream to maintain synchronization. In response to the transmission of SYNC characters, the Transmitter Empty (TxEMPTY) line is raised high. The CPU can also be made aware of this condition by reading bit D2 of the Status Register (also TxEMPTY). The TxEMPTY line and Status bit are reset when the microprocessor sends new character data to the 8251.
- *Data Reception.* The 8251 must be synchronized with the device sending data in order to perform as a synchronous receiver. The USART must also be enabled for receive operation by setting Command Instruction bit D2. Synchronization can be achieved either by internal or external means. Internal SYNC mode is selected by setting bit D7 of the Command Instruction to one. This puts the USART into the ENTER HUNT mode of operation. As a part of synchronous operation, one or two SYNC characters are programmed into the 8251 immediately after the Mode Instruction. In ENTER HUNT mode the USART compares the programmed SYNC characters (one or two characters are possible) to the information received from the sending device. When the received and programmed SYNC characters match, synchronization is achieved and the SYNDET output line goes high. This line is reset by reading the Status Register. External SYNC also uses the SYNDET line but now as an input

(SYNDET is configured using the Mode Instruction). Synchronization is obtained by holding this line high for a full Receiver Clock cycle.

Once character synchronization is obtained, data is received at a rate equal to the Receiver Clock. The Receiver Ready line (RxRDY) goes high to indicate that a received character is available for the microprocessor. Parity and overrun errors are detected in synchronous mode.

If, for some reason synchronization is lost, another Command Instruction can specify that HUNT Mode be reentered.

USART Instructions, Registers, and Modem Control Lines

One final bit in the Command Instruction format needs to be mentioned. Bit D3 is the SEND BREAK CHARACTER. When activated (high level), the Transmitter output line is forced to a continuous low level until a subsequent Command Instruction deactivates the SEND BREAK CHARACTER bit. This allows the Transmitter to send a break character whose duration is longer than the time for a full character (i.e., all zeros sent as data would not be detected by a receiving device as break characters, since the data would be framed by start, stop, and parity bits. Consequently, the receiver would not see constant zeros for a sufficient length of time for them to be perceived as break characters.) When the USART is receiving data in asynchronous mode, the SYNDET/BRKDET line will go high if a break character has been received through two full character time sequences.

The Status Register keeps track of the USART interval activity and also provides the means for the microprocessor to use the USART in a polled environment rather than an interrupt environment. Many of the external pins on the USART also have, in the Status Register, corresponding bits that can be read by the microprocessor (TxRDY is an exception). Figure II.4.6 illustrates the Status Register format. Through software, the CPU can read the Status Register and obtain information on the internal condition of the USART.

Four lines are provided for modem control on the 8251. The two lines Data Set Ready ($\overline{\text{DSR}}$) and Clear To Send ($\overline{\text{CTS}}$) are inputs to the USART from the modem. The Data Set Ready line informs the USART that the modem has established a call connection. Clear To Send is the modem's response that a data transmission from the USART may begin. Two output lines are also available for modem control, Data Ter-

Figure II.4.6 Status Register

Courtesy of Intel Corp.

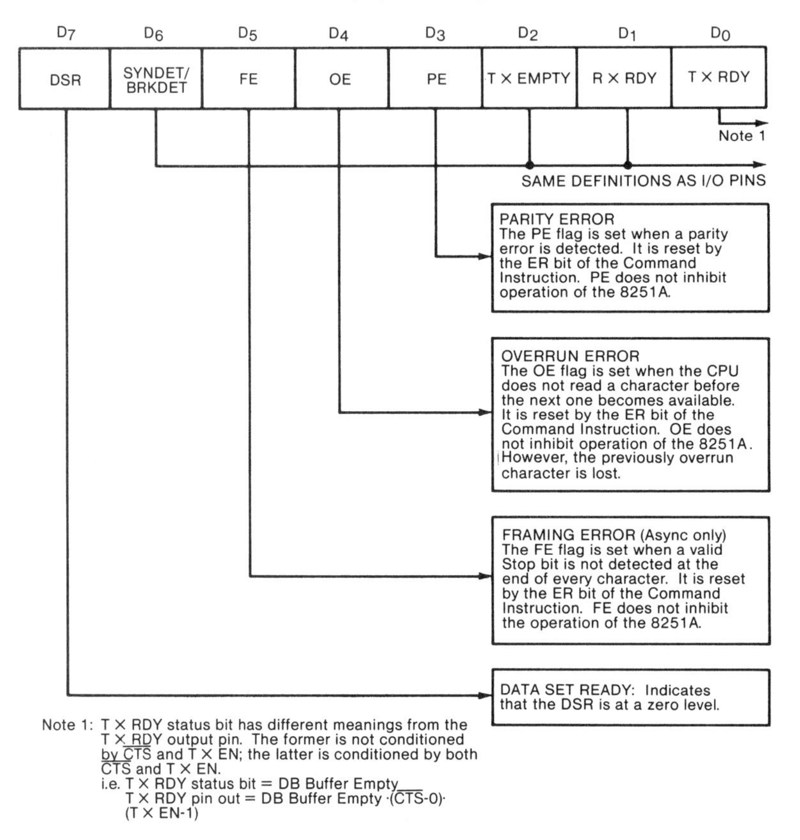

minal Ready ($\overline{DTR}$) and Request To Send ($\overline{RTS}$). When Request To Send is lowered (via a Command Instruction), the 8251 informs the modem to activate its carrier in preparation for a data transmission. Data Terminal Ready is the USART's indication that incoming data can be accepted (also made active by issuing a Command Instruction).

II.5

Analog-to-Digital Converter

PART NUMBER	AD574A
FUNCTION	ANALOG-TO-DIGITAL CONVERTER
MANUFACTURER	ANALOG DEVICES
VOLTAGES	$V_{LOGIC} = +5$ $Vcc = +12/+15$ $Vee = -12/-15$
PWR. DISS.	390 mW
PACKAGE	28-pin DIP
TEMPERATURE	0°C → +70°C −55°C → + 125°C
FEATURES	12-BIT A/D CONVERTER BUILT-IN VOLTAGE REFERENCE 8- OR 16-BIT MICROPROCESSOR INTERFACE 250 nsec BUS ACCESS TIME FOUR CALIBRATED RANGES UNIPOLAR AND BIPOLAR OPERATION
COMPATIBLE MICROPROCESSORS	MOST 8- AND 16-BIT PROCESSORS
FUNCTIONAL DESCRIPTION	THE SUCCESSIVE APPROXIMATION TECHNIQUE IS USED TO IMPLEMENT THE CONVERSION PROCESS WITHIN THIS 12-BIT A/D CONVERTER. THREE-STATE OUTPUTS ALLOW A DIRECT INTERFACE TO 8- AND 16-BIT PROCESSORS. 12-BIT OUTPUT DATA CAN BE OBTAINED IN A SINGLE 16-BIT OR TWO 8-BIT TRANSFERS. AN EXTERNAL RESISTORS NETWORK SETS USABLE ANALOG VOLTAGE RANGES.

Figure II.5.1 AD574A Block Diagram and Pinout

Courtesy of Analog Devices, Inc.

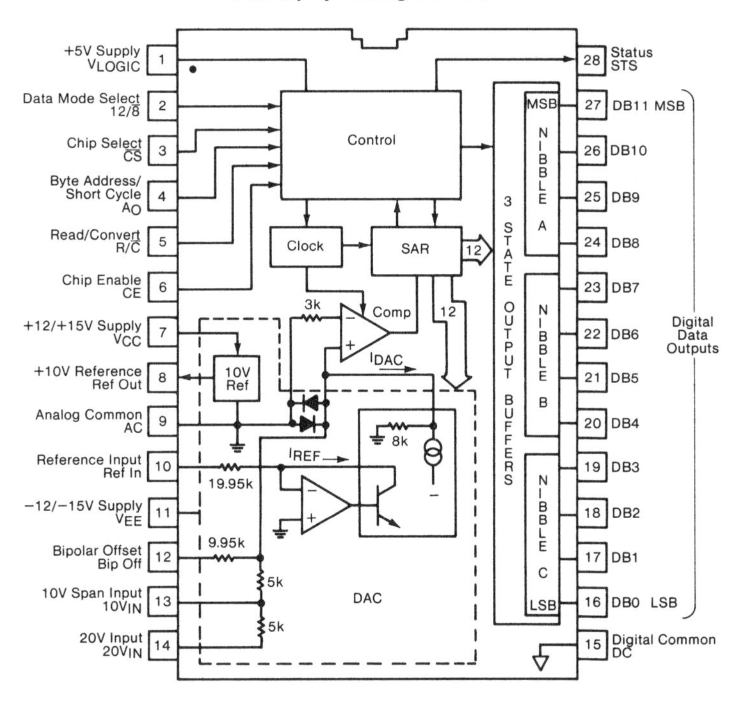

Figure II.5.2 Unipolar Input Connections

Courtesy of Analog Devices, Inc.

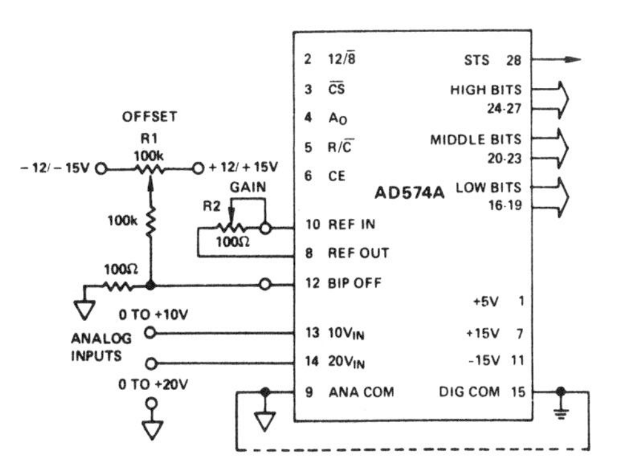

PIN NAME	PIN SYMBOL	PIN NUMBER	FUNCTION
CHIP SELECT	$\overline{CS}$	3	Chip select, when low, enables the A/D Converter for both read and conversion processes.
CHIP ENABLE	CE	6	Chip enable, when high, enables the A/D Converter for both read and conversion processes.
READ/CONVERT	$R/\overline{C}$	5	The state of the $R/\overline{C}$ lines determines whether a read or conversion process will take place. CE and $\overline{CS}$ also must be active for these processes to be initiated.
DATA MODE SELECT	$12/\overline{8}$	2	This line aids in the interfacing of the A/D converter to 8-bit and 16-bit microprocessor buses. When tied to +5 V, the digital data resulting from an A/D conversion is placed on the Digital Data Outputs (DB0 - DB11) as a 12-bit quantity. When tied to ground, the digital data is presented as two separate 8-bit quantities.
BYTE ADDRESS/ SHORT CYCLE	A0	4	When low, this line will cause a 12-bit A/D conversion to take place. A high level produces a shorter 8-bit conversion. This line is usually controlled by the least significant bit of the system address bus.

PIN NAME	PIN SYMBOL	PIN NUMBER	FUNCTION
STATUS	STS	28	Status, when high, indicates that a conversion process has begun. The line will return low when the conversion is complete.
DATA BUS 0	DB0	16	These 12 output lines provide the digital data that results from the A/D conversion process. When inactive, the lines are placed in a high impedance state.
DATA BUS 1	DB1	17	
DATA BUS 2	DB2	18	
DATA BUS 3	DB3	19	
DATA BUS 4	DB4	20	
DATA BUS 5	DB5	21	
DATA BUS 6	DB6	22	
DATA BUS 7	DB7	23	
DATA BUS 8	DB8	24	
DATA BUS 9	DB9	25	
DATA BUS 10	DB10	26	
DATA BUS 11	DB11	27	
+5-V SUPPLY	V_{LOGIC}	1	The 5-V power supply input is used to establish proper logic levels.
DIGITAL COMMON	DC	15	This ground point for digital signals is typically connected to the Analog Common pin (9).
10-V SPAN INPUT	$10V_{IN}$	13	This is one of the analog signal inputs. This input is designed for analog signals that reach a maximum level of 10 V. The input may be used in a bipolar or unipolar mode of operation.
20-V SPAN INPUT	$20V_{IN}$	14	This is the second of two analog signal inputs. This

(*continued*)

PIN NAME	PIN SYMBOL	PIN NUMBER	FUNCTION
			input is designed for analog signals that reach a maximum level of 20 V. The input may be used in a bipolar or unipolar mode of operation.
BIPOLAR OFFSET	BIP OFF	12	Used with external resistors, this pin is used to provide an offset trim adjustment for improved accuracy. The pin is grounded if not needed.
+12/+15-V SUPPLY	VCC	7	A supply of +12 or +15 V is connected to this pin to provide power for the analog section of the chip.
−12/−15-V SUPPLY	VEE	11	A supply of −12 or −15 V is connected to this pin to provide power for the analog section of the chip.
+10-V REFERENCE	REF OUT	8	A 10-V reference voltage is provided at this output. This line may be used to drive a reference input resistor, a bipolar offset resistor, and other external loads. Drive capability for external loads is 1.5 mA.
REFERENCE INPUT	REF IN	10	The Reference Input is usually obtained from an external resistor powered by the 10-V Reference output. This input is a part of the trim adjustments made

PIN NAME	PIN SYMBOL	PIN NUMBER	FUNCTION
			to the chip for improved accuracy.
ANALOG COMMON	AC	9	This pin is used as the internal ground reference for the A/D chip and should be connected to the same ground reference point as that of the analog system being tested.

While the analog-to-digital conversion process is often carried out without the aid of a microprocessor, the very fact that the output of this process is a digital word encourages the joining of the A/D system to a microprocessor system. The precision of a well-designed A/D chip along with the storage and decision making capacity of the microprocessor can result in a data acquisition system of powerful sophistication and versatility.

AD574A BASIC OPERATION

The AD574A (Fig. II.5.1) is a 12-bit, successive-approximation A/D Converter. An analog input voltage is converted into a 12-bit digital word upon command from the microprocessor. After the conversion is complete, the digital word is placed on the system data bus for CPU access and processing. To accommodate both 8- and 16-bit microprocessors, the data can be presented to the data bus in one of two modes. One chip pin ($12/\overline{8}$) is used to distinguish the modes and allows data output as one 12-bit data word or two 8-bit data words.

The conversion process begins when the CPU asserts the Read/Conversion ($R/\overline{C}$) control line and selects the chip, through appropriate addressing means (CE and $\overline{CS}$). When the A/D conversion is completed, a Status signal (STS) informs the CPU that converted data is ready. The CPU can then read this data by raising the $R/\overline{C}$ line.

From an analog standpoint, voltages between 0 to +10, 0 to +20, −5 to +5, or −10 to +10 can be converted with this chip. A small number of external resistors are used to configure the chip for a specific range of input voltages as well as to provide offset control. Three power supplies are necessary for proper chip operation. A +5 V supply powers the logic sections of the chip while supplies of +12/+15 V and −12/−15 V take care of the analog sections.

THE ANALOG-TO-DIGITAL CONVERSION PROCESS

The digital word that is output from the A/D conversion process is originally developed within an internal 12-bit Successive-Approximation Register (SAR). This register, in conjunction with an internal Digital-to-Analog Converter (DAC), a comparator, a clock, and an associated control circuit, can convert an analog voltage into an equivalent digital code in a very short period of time. Basically, the SAR is sequenced, one bit at a time, from the MSB to the LSB. The bits are individually tested to see if their binary weight would bring the resulting digital value closer to the analog value. This is accomplished by taking the SAR value and running it through the on-board DAC. The DAC will produce a temporary analog value that is compared to the actual analog voltage at the chip's input. Since two analog voltages are now under comparison, the on chip hardware can now determine whether the SAR value is acceptable, too high or too low. The SAR is modified as appropriate, with the process continuing through all twelve bits. The final value in the SAR is then the correct digital code representing the value of the analog voltage.

Two inputs are available for analog voltages. The choice of input used, along with other external connections, gives the chip the capability of accepting analog voltages within four different ranges. Two basic types of connections are possible, unipolar and bipolar.

Configuring the Chip for Unipolar Operation

A unipolar connection sets up the A/D Converter to accept positive voltages only in the ranges of 0 to +10 V or 0 to +20 V. Voltages in the 0- to +10-V range are presented to pin 13, the 10-V Span Input, while voltages in the other range are connected to pin 14, the 20-V Span Input. It is evident that the input used depends upon the application at hand. Pin 9 (Analog Common) is the analog voltage reference point. In its simplest form, an analog input voltage and an analog ground are virtually all that is necessary for a functional system. The designer must, however, consider the accuracy of the resultant digital information with this basic system. Under these circumstances, maximum zero offset error is ± 2 LSB (1 LSB is the smallest incremental change that the A/D Converter can make) and maximum full-scale error is 10 LSB for the AD574AK. This means that the digital value for an analog input of zero volts could be off by as much as the voltage equivalent of 2 bits and that the maximum digital output (all ones) could occur with an input voltage that is actually a voltage 10 bits lower than it should be. These specifications, of course, will translate into voltage offsets that can be calculated once the input voltage range is known. For instance, assume that voltages in the range of 0 to +10 V are being used. This converter is a 12-bit converter. Therefore, the 10-V voltage range will be divided up into 4096 steps ($2^{12} = 4096$). Ten volts divided by the number of steps gives the value of the Least Significant Bit or 10/4096 = 2.44 mV. Therefore, a change in the LSB by one is equivalent to the analog input voltage changing by 2.44 mV. A digital value of all zeros could be obtained from an input voltage of 4.88 mV, resulting in an error, since the maximum error can be off by as much as 2 LSB (2.44 mV × 2 = 4.88 mV). The full-scale output could be off by as much as 24.4 mV. If these variations from the nominal values are satisfactory for the application at hand, then the basic system can be used. If greater accuracy is necessary, then offset voltages can be introduced to balance the errors.

As shown in Figure II.5.2, offset control is obtained for both full-scale and zero readings by connecting resistors to pins designed for this purpose. The Bipolar Offset input (pin 12), when connected as shown, will take care of the full-scale error. To calibrate the system, an analog input voltage equal to the full-scale voltage, (defined as being 1.5 LSB below the nominal full-scale voltage), is applied to the analog input. For the 10-V input in our example, this voltage would equal 9.99634 V (10 − 1.5 × 2.44 mV). The variable resistor, R1, is then adjusted so that the digital output changes from 111111111110 to 111111111111 with this input voltage. (A digital representation for 10 V does not exist in this example. An output of all ones occurs for a voltage equal to 9.99634 V.)

Pin 8 is an output from the internal voltage reference circuitry and is used along with pin 10, a reference input pin, to create a zero offset. This A/D Converter is designed so that the analog voltage for any equivalent digital code occurs halfway between the codes. For example, using a 10-V system, the change from 000000000000 to 000000000001 should occur for a voltage of 1.22 mV. Recall that the 10-V analog range was divided into 4096 pieces of 2.44 mV each. A voltage of 1.22

mV is exactly one-half of this value and is the voltage input that should be used to calibrate the zero offset. Adjusting resistor R2 in Figure II.5.2 will trim the zero offset.

With these offset resistors, both the high and low voltage values can be precisely set, resulting in an extremely accurate A/D system. It should be noted that for the basic system without offset capability, pin 12 is connected to Analog Common and a 50 ohm 1% metal film resistor is connected between pins 8 and 10.

Configuring the Chip for Bipolar Operation

An analog voltage range, varying between negative and positive limits, is acceptable to the AD574A using a bipolar connection. Figure II.5.3 clearly illustrates that input voltages within the range −5 V to +5 V are accepted by the chip at pin 13 while the larger voltage span, −10 V to +10 V, is input at pin 14. The Analog Common pin is used as the ground reference in both cases. All the pins discussed previously, with the exception of the Bipolar Offset and analog input pins, are treated the same.

The offsets are not necessary unless extreme accuracy is required. If this is the case, the trimming resistors, R1 and R2, can be replaced by a 50-ohm resistor. If offset trimmers are used, then calibration is necessary. The calibration procedure is similar to that of the unipolar connection. The zero offset (R1) is adjusted so that an input voltage of 0.5 LSB above the maximum negative voltage gives a digital output code that changes from 000000000000 to 000000000001. For the span,

Figure II.5.3 Bipolar Input Connections

Courtesy of Analog Devices, Inc.

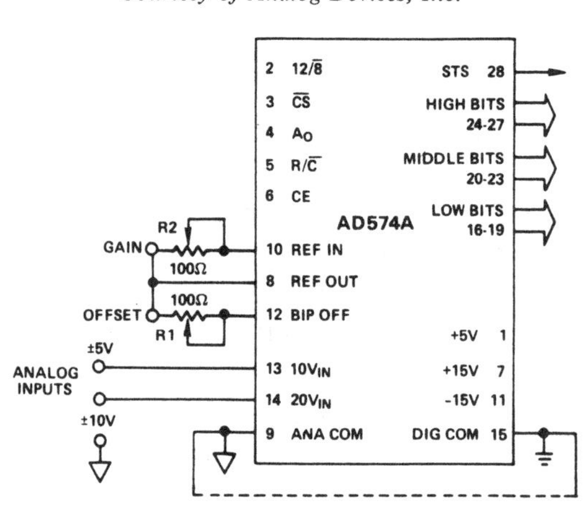

−5 to +5 V, this means that this change in digital value occurs for an input voltage of −4.9988 V. The full-scale offset trim (R2) is adjusted for an input 1.5 LSB below maximum positive full-scale voltage. For the 5-V range, an input of +4.9963 V will cause the digital code to change from 111111111110 to 111111111111.

THE MICROPROCESSOR INTERFACE

This A/D Converter can interface to most microprocessors using the common control pins provided. Two select inputs, Chip Enable (CE) and Chip Select ($\overline{CS}$), are both activated to begin processing with the AD574A. Either control line can initiate chip activity, assuming that the other is already activated. The only difference between the two is access time (CS is 100 nsec longer).

Once the chip is enabled, the Read/Control (R/$\overline{C}$) determines the mode of operation. A conversion process is begun by selecting the chip and asserting R/$\overline{C}$ low. The Status (STS) output then goes high for the duration of the analog-to-digital conversion process and returns low when the conversion process is complete. The Status line may or may not be used by the microprocessor. For example, the conversion time for this chip is 35 μsec maximum. Once a conversion has begun (via CPU control), the CPU can execute other instructions and then return to the A/D Converter to read data, knowing that after 35 μsec have elapsed output data is ready. In this case the Status line is unnecessary. Alternatively, the Status line could be used to interrupt the processor when the conversion process has ended. Interrupts would generally give the CPU a faster response to the converted data.

Two conversions, an 8-bit or 12-bit, are selectable by proper addressing. The state of address bit A0, when a conversion process is started, determines the mode of operation. If A0 is low, then a 12-bit conversion cycle is started. When A0 is high, an 8-bit conversion cycle takes place. Either conversion mode can be used, but resolution is reduced in 8-bit mode.

After the analog voltage has been converted into a digital word, the CPU must read the AD574A to obtain the digital information. $\overline{CS}$ and CE are activated along with R/$\overline{C}$ set high. These signals will enable the chip and cause data to be placed on the data bus. The manner in which data is accessed by the CPU depends on several things. First, the size of the data bus will dictate whether one or two read operations are needed to completely acquire the digital data. A 16-bit data bus is specified by tying the 12/$\overline{8}$ pin to V_{Logic}. The designer can then connect

Figure II.5.4 AD574A Data Format for 8-Bit Bus

Courtesy of Analog Devices, Inc.

	D7							D0
XXX0 (EVEN ADDR):	DB11 (MSB)	DB10	DB9	DB8	DB7	DB6	DB5	DB4
XXX1 (ODD ADDR):	DB3	DB2	DB1	DB0 (LSB)	0	0	0	0

all 12 data output pins to the 16-bit microprocessor data bus in any suitable fashion (i.e., starting with the LSB or MSB). The four unused data bus pins will mean that a masking operation in software will be required, since data on these pins will be meaningless. When $12/\overline{8}$ is high, all 12 data bits become available when a read ($R/\overline{C} = 1$) is begun (assuming that a conversion is not in process).

When $12/\overline{8}$ is tied to Digital Common, an 8-bit data bus interface is specified. Two read operations will be needed to recover the 12-bit digital information. A0 specifies exactly how the data is retrieved. With A0 = low, the eight most significant bits of the digital information are output during the read. If A0 is high, then the remaining four least significant bits are output as part of an 8-bit data byte and organized so that the four data bits are followed by four zeros (Fig. II.5.4). The four zeros eliminate the need for software masking.

Access times for this chip are equal to those of common memory devices, making a direct connection to the microprocessor bus possible. Once the CPU has obtained the digital equivalent of the analog input voltage, a software program is used to process the data.

II.6

Digital-to-Analog Converter

PART NUMBER	DAC-888
FUNCTION	8-BIT MULTIPLYING D/A CONVERTER
MANUFACTURER	PRECISION MONOLITHICS INC.

VOLTAGES	+V TO −V = 18.1 V	**PWR. DISS.**	300 mW MAX
PACKAGE	18-pin DIP	**TEMPERATURE**	−25°C → +85°C

FEATURES

8-BIT LEVEL TRIGGERED LATCH, TTL COMPATIBLE
8-BIT D/A, ±0.19% NONLINEARITY
HIGH OUTPUT IMPEDANCE
COMPLEMENTARY ANALOG OUTPUTS
VARIABLE INPUT REFERENCE CURRENT
BUFFERED INPUTS

COMPATIBLE MICROPROCESSORS: ALL 8-BIT MICROPROCESSORS

FUNCTIONAL DESCRIPTION

THIS 8-BIT MULTIPLYING DAC WILL PRODUCE AN OUTPUT CURRENT THAT IS THE PRODUCT OF THE INPUT 8-BIT DIGITAL WORD AND AN INPUT REFERENCE CURRENT. THE INPUT DATA IS LATCHED WITHIN THE CHIP UNDER CONTROL OF CHIP ENABLE AND WRITE CONTROL LINES. TYPICAL SETTLING TIME IS 300 nsec.

Figure II.6.1 DAC-888 Pinout

Courtesy of Precision Monolithics, Inc.

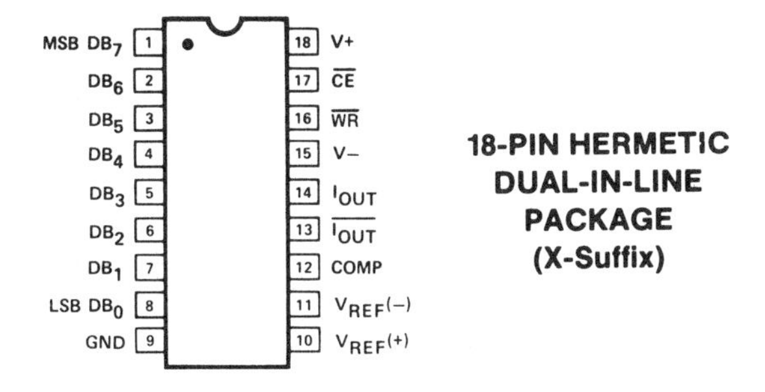

Figure II.6.2 DAC-888 Block Diagram

Courtesy of Precision Monolithics, Inc.

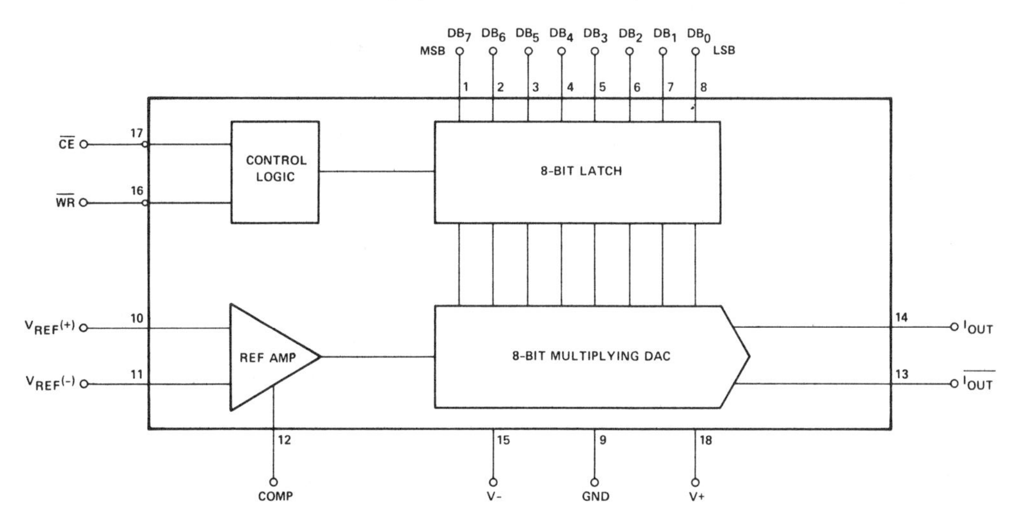

PIN NAME	PIN SYMBOL	PIN NUMBER	FUNCTION
CHIP ENABLE	$\overline{CE}$	17	When low and in conjunction with $\overline{WR}$, $\overline{CE}$ enables the DAC-888 for a data write operation.
WRITE CONTROL	$\overline{WR}$	16	When low and in conjunction with $\overline{CE}$, $\overline{WR}$ controls a data write operation to the DAC-888. Data written to the chip is subsequently converted into analog form.
DATA BIT 0 **DATA BIT 1** **DATA BIT 2** **DATA BIT 3** **DATA BIT 4** **DATA BIT 5** **DATA BIT 6** **DATA BIT 7**	DB0 DB1 DB2 DB3 DB4 DB5 DB6 DB7	8 7 6 5 4 3 2 1	Data information from the microprocessor is presented to the DAC on these lines. The setup time for data is 100 nsec prior to $\overline{CE}$ and $\overline{WR}$ becoming active.
VOLTAGE REFERENCE+	V_{REF+}	10	This positive voltage input terminal is used to set a positive reference current flow.
VOLTAGE REFERENCE−	V_{REF-}	11	This negative voltage input terminal is used to set a negative reference current flow.
CURRENT OUTPUT **CURRENT OUTPUT**	I_{OUT} $\overline{I_{OUT}}$	14 13	These complementary outputs provide a current flow representing the value of the input digital word.
COMPENSATION	COMP	12	A frequency compensating capacitor connected from

PIN NAME	PIN SYMBOL	PIN NUMBER	FUNCTION
			this pin to ground stabilizes the internal reference amplifier.
V+	V+	18	Positive supply voltage.
V−	V−	15	Negative supply voltage.
GROUND	GND	9	Ground.

Converting binary logic information into a corresponding analog representation is a common process. If the source of binary information is a microprocessor, or more specifically, the data bus of a microprocessor, then several design elements need to be considered before a valid digital-to-analog conversion can begin. First, the voltage levels of the input binary data must be compatible with the converting device. Since most microprocessors convey their information in the form of TTL logic levels, the D/A converter must be able to process the voltage variations within the valid logic one and logic zero ranges. Second, the D/A converter should present as small a load as possible to the microprocessor's data bus. Therefore, a buffered input is desirable to minimize the effects on system fan-out. A third consideration concerns the transient nature of the data bus contents. Since data is present for only a very short period of time, data latches are required to store the data information for its eventual conversion into analog form. Finally, microprocessor control lines that synchronize the transmission of correct data to the D/A converter are necessary.

The DAC-888 (Figs. II.6.1 and II.6.2) was designed to meet the requirements of a digital-to-analog conversion under microprocessor control.

THE DIGITAL CONNECTION

The 8-bit-wide, buffered data path of the DAC-888 makes this device ideal for use with 8-bit microprocessors. Binary data destined to undergo a digital-to-analog conversion is placed onto the microprocessor's data bus by system software. This is accomplished either through a memory write operation or through an I/O write operation. Common CPUs on the market today include a control line that is asserted when the CPU engages in a write cycle. The DAC-888 has an input pin ($\overline{WR}$) designated to sense this occurrence. An additional Chip Enable ($\overline{CE}$) pin is also used to synchronize the transfer of data between CPU and DAC. The $\overline{CE}$ pin is typically driven low by external address bus decoding hardware and, when active, enables the DAC for the data transfer. Data is internally latched when both $\overline{CE}$ and $\overline{WR}$ are activated. The data needs to be available only 100 nsec before one of the two control lines is deactivated (Setup Time). Once the data is latched within the DAC, the control lines may be deactivated. The digital-to-analog conversion will then take place independent of any CPU involvement. The analog output is available 300 nsec after Chip Enable is made inactive (Settling Time). Figure II.6.3 shows the DAC interface to a 6502 microprocessor.

THE ANALOG CONNECTION

The analog output of the DAC-888 is a current flow that increases or decreases proportionally as the binary word changes value. Two outputs are available that are analog complements of one another. In all cases, the sum of I_{out} and $\overline{I_{out}}$ will equal the Full Range Output Current (I_{FR}). I_{FR} is the maximum value of output current available at either output pin. Figure II.6.4 illustrates the values of output current for corresponding values of digital input. An analog output voltage is easily obtained from the output current by sensing the voltage drop created across a resistor. In this example, the Full Range Output Current is equal to 1.992 mA.

I_{FR} is found by the following formula:

$$I_{FR} = 255/256 \times I_{REF}$$

For the example shown, the value of I_{REF} is given as 2 mA. Therefore, $I_{FR} = 255/256 \times 2\text{ mA} = 1.992\text{ mA}$. This current level is found at the I_{out} output for a digital input of 11111111. Notice that the $\overline{I_{out}}$ output is equal to 0.0 mA under these circumstances. If one of the outputs is not used it must be tied either to ground or to a point able to handle the Full Range Output Current. I_{out} and $\overline{I_{out}}$ should not be left open.

Figure II.6.3 6502 Interface

Courtesy of Precision Monolithics, Inc.

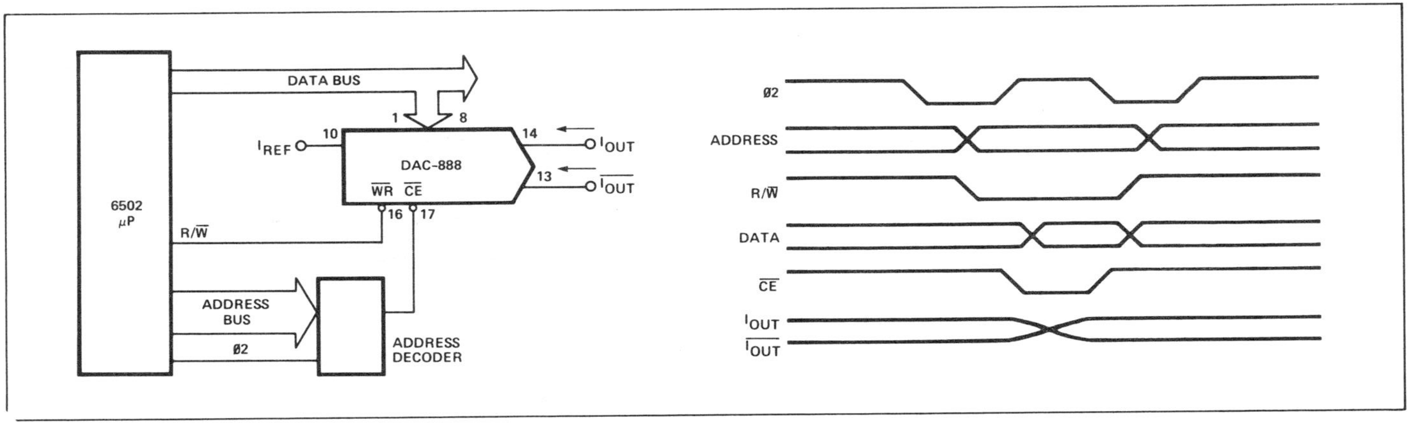

Figure II.6.4 Basic Unipolar Negative Operation

Courtesy of Precision Monolithics, Inc.

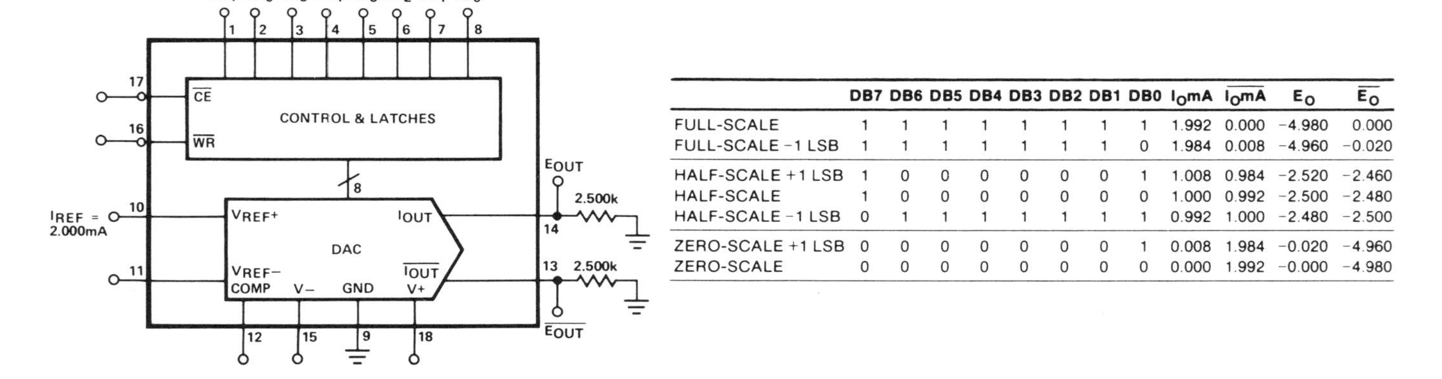

	DB7	DB6	DB5	DB4	DB3	DB2	DB1	DB0	I_OmA	$\overline{I_O}$mA	E_O	$\overline{E_O}$
FULL-SCALE	1	1	1	1	1	1	1	1	1.992	0.000	−4.980	0.000
FULL-SCALE −1 LSB	1	1	1	1	1	1	1	0	1.984	0.008	−4.960	−0.020
HALF-SCALE +1 LSB	1	0	0	0	0	0	0	1	1.008	0.984	−2.520	−2.460
HALF-SCALE	1	0	0	0	0	0	0	0	1.000	0.992	−2.500	−2.480
HALF-SCALE −1 LSB	0	1	1	1	1	1	1	1	0.992	1.000	−2.480	−2.500
ZERO-SCALE +1 LSB	0	0	0	0	0	0	0	1	0.008	1.984	−0.020	−4.960
ZERO-SCALE	0	0	0	0	0	0	0	0	0.000	1.992	−0.000	−4.980

HOW TO SET THE REFERENCE CURRENT

The actual value of analog current flow is equal to the digital input value times the input reference current. Specifically, I_{out} = (Digital Input Value/256) × I_{REF}. The complement output $\overline{I_{out}}$, can be found by: $\overline{I_{out}} = I_{FR} - I_{out}$.

Setting an accurate level of input reference current is critical to obtaining the desired levels of analog output current. Since this is a multiplying DAC, the input reference current level may be varied, resulting in output currents that are multiples of the input reference. For instance, an input reference current set at 1.0039 mA will result in an output current of 1 mA for a digital input of 11111111 (I_{out} = 255/256 × 1.0039 mA = 1 mA). If the input reference current were set to 2.0078 mA, then the same digital input of 11111111 would cause an output current level of 2 mA (I_{out} = 255/256 × 2.0078 mA = 2 mA).

The recommended range of input reference current is +0.2 mA to +4.0 mA. The reference current can be set for positive, negative, or bipolar operation by assiduous control of reference input pins 10 and 11. An external voltage reference connected to pin 10 through a resistor sets the level of I_{FR}. For TTL operation, a resistor value of 2.5K and a voltage reference level of 5 V is typical. (The manufacturer does

Figure II.6.5 Basic Positive Reference Operation
Courtesy of Precision Monolithics, Inc.

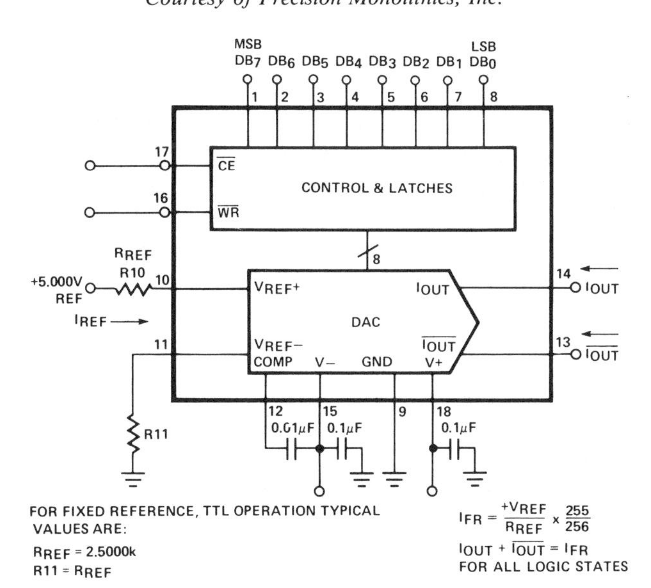

Figure II.6.6 Basic Bipolar Output Operation

Courtesy of Precision Monolithics, Inc.

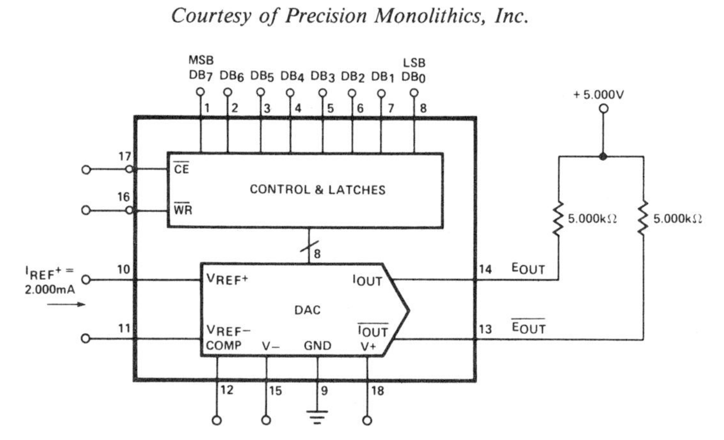

	DB7	DB6	DB5	DB4	DB3	DB2	DB1	DB0	E_O	$\overline{E_O}$
POSITIVE FULL-SCALE	1	1	1	1	1	1	1	1	−4.960	5.000
POSITIVE FULL-SCALE −1 LSB	1	1	1	1	1	1	1	0	−4.920	4.960
ZERO-SCALE +1LSB	1	0	0	0	0	0	0	1	−0.040	0.080
ZERO-SCALE	1	0	0	0	0	0	0	0	0.000	0.040
ZERO-SCALE −1LSB	0	1	1	1	1	1	1	1	0.040	0.000
NEGATIVE FULL-SCALE +1 LSB	0	0	0	0	0	0	0	1	4.900	−4.920
NEGATIVE FULL-SCALE	0	0	0	0	0	0	0	0	5.000	−4.960

not recommend using the TTL supply for the 5-V reference voltage.) Figure II.6.5 indicates the proper connections for a positive reference current. The resistor connected to pin 11 would be the same value as the reference resistor in order to minimize bias current errors.

A negative current reference is set in a similar fashion by connecting a negative voltage reference through a resistor to pin 11. In this case, pin 10 is connected to ground through a resistor. Output current flow has the opposite polarity from that in the positive reference case.

Bipolar operation is also possible as shown in Figure II.6.6. This connection splits the range of digital input values into positive and negative output values.

AC input references may also be used but require compensating capacitors. The actual capacitor value depends upon the impedance of the AC driving source. In general, lower values of resistance enable the use of lower value capacitors.

II.7

Universal Interrupt Controller

PART NUMBER	Am9519A
FUNCTION	UNIVERSAL INTERRUPT CONTROLLER
MANUFACTURER	ADVANCED MICRO DEVICES
VOLTAGES	Vcc = +5
PWR. DISS.	400 mW
PACKAGE	28-pin DIP 28-pin LCC
TEMPERATURE	0°C → +70°C
FEATURES	EIGHT MASKABLE INTERRUPTS, EXPANDABLE POLLED OR VECTOR INTERRUPT OPERATION ROTATING AND FIXED PRIORITY LOGIC PROGRAMMABLE VECTOR MEMORY
COMPATIBLE MICROPROCESSORS	MOST 8- AND 16-BIT PROCESSORS
FUNCTIONAL DESCRIPTION	THIS INTERRUPT CONTROLLER EXPANDS THE INTERRUPT CAPABILITY AND FLEXIBILITY OF A MICROPROCESSOR. EIGHT MASKABLE INTERRUPTS ARE SUPPORTED BY THE CHIP. THE PRIORITY OF THE INTERRUPTS CAN BE SET FOR VARIOUS APPLICATIONS. AN INTERRUPT IS RESOLVED BY THE CONTROLLER FOR PRIORITY; THEN ITS ASSOCIATED VECTOR IS PLACED ON THE DATA BUS FOR CPU SERVICE. PROGRAMMABILITY AND AUTOMATIC HARDWARE CLEAR ALLOW FOR AN INTERRUPT STRUCTURE WITH A MINIMUM SOFTWARE BURDEN.

Figure II.7.1 Am9519A Pinout

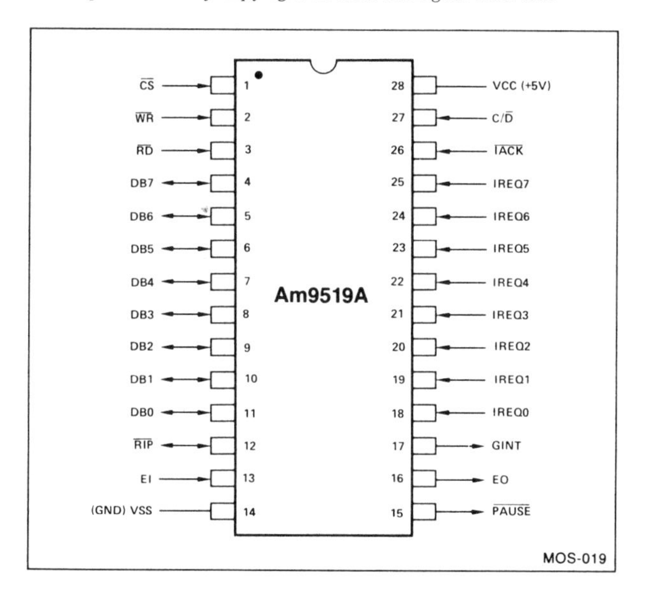

Figure II.7.2 Am9519A Block Diagram

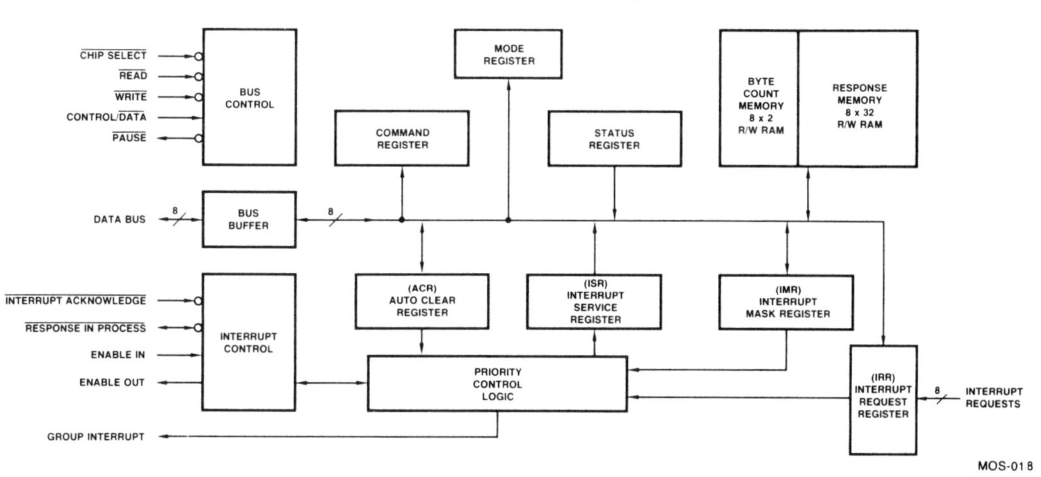

PIN NAME	PIN SYMBOL	PIN NUMBER	FUNCTION
CHIP SELECT	$\overline{CS}$	1	Commands and initialization information may be transferred between the interrupt controller and the CPU when $\overline{CS}$ is low. Interrupts will take place regardless of the level on $\overline{CS}$.
WRITE READ	$\overline{WR}$ $\overline{RD}$	2 3	When low (only one line can be low at a time), these two CPU control lines enable the Am9519A to receive data from or send data to the CPU. This exchange of data takes place over the data bus.
CONTROL/DATA	C/$\overline{D}$	27	Control/$\overline{Data}$, in conjunction with $\overline{RD}$, $\overline{WR}$, and $\overline{CS}$, determine, in part, the registers or memory locations for data transfers.
DATA BUS 0 DATA BUS 1 DATA BUS 2 DATA BUS 3 DATA BUS 4 DATA BUS 5 DATA BUS 6 DATA BUS 7	DB0 DB1 DB2 DB3 DB4 DB5 DB6 DB7	11 10 9 8 7 6 5 4	All programming commands and data are exchanged between the interrupt controller and the CPU over these lines.
INTERRUPT REQUEST 0 INTERRUPT REQUEST 1 INTERRUPT REQUEST 2	IREQ0 IREQ1 IREQ2	18 19 20	An interrupting device signals its need for an interrupt over these lines. The state of the interrupt request line is programmable for either a positive or

PIN NAME	PIN SYMBOL	PIN NUMBER	FUNCTION
INTERRUPT REQUEST 3	IREQ3	21	negative edge, and is latched internally when active.
INTERRUPT REQUEST 4	IREQ4	22	
INTERRUPT REQUEST 5	IREQ5	23	
INTERRUPT REQUEST 6	IREQ6	24	
INTERRUPT REQUEST 7	IREQ7	25	
GROUP INTERRUPT	GINT	17	This output, which can be programmed for a high or low active level, indicates to the CPU that an interrupt is pending. This pin is not suitable for edge-sensitive devices and requires an external pull-up for the active low response.
INTERRUPT ACKNOWLEDGE	$\overline{IACK}$	26	This low active input is asserted by the CPU during the interrupt process to signal the need for interrupt response information. The Am9519A can be programmed to respond to a varied number of $\overline{IACK}$ pulses (1–4).
RESPONSE IN PROGRESS	$\overline{RIP}$	12	This bidirectional signal is used when two or more Am9519As are used in cascade. $\overline{RIP}$ allows one chip to complete its interrupt activity without interference from other interrupt controller chips.

(continued)

PIN NAME	PIN SYMBOL	PIN NUMBER	FUNCTION
PAUSE	$\overline{\text{PAUSE}}$	15	$\overline{\text{PAUSE}}$ is an active low line (pull-up required) used to adjust the timing differences between the data bus and control lines due to the delays in priority resolution logic.
ENABLE IN	EI	13	This input is connected to the Enable Out (EO) output of other interrupt controllers in a daisy-chain cascade. The connection of these pins determines the priority sequence of several controllers in the same interrupt system. EI, when low, will disable the interrupt system.
ENABLE OUT	EO	16	Used with the EI pin, EO is pulled low if an interrupt request is present on chip. The low on EO will disable all lower priority chips.
Vcc	Vcc	28	+5 V.
GROUND	Vss	14	Ground.

An interrupt structure is a powerful feature in any microprocessor system. Succeeding generations of microprocessors have been upgraded to make the implementation of interrupts easier for the software and hardware designer. Options such as vectoring and masking lead to these powerful structures. The limited number of CPU pins available, however, also limits the extent of CPU hardware support for interrupts. Therefore, peripheral chips are often used for large interrupt systems. The Am9519A is such a chip. Many outstanding features designed into the Am9519A can be employed by the designer to create a sophisticated interrupting system.

UNDERSTANDING INTERRUPT CONTROLLER OPERATION

Each Am9519A can respond to eight interrupting devices. By cascading Interrupt Controller chips together, an expanded interrupt structure can be developed. In many instances, some interrupting devices will take precedence over others. Several priority schemes designed into the controller chip will fit applications where specific priorities are a necessity.

Figures II.7.1 and II.7.2 illustrate the pinout and block diagram of the Am9519A. Registers are used to hold the state of interrupts, mask interrupts, store commands, and determine modes of operation, as well as to perform other tasks. A software initialization routine is run to configure the chip for a particular mode of operation. This programming can be altered during the course of normal CPU activity or during an interrupt as the situation dictates. For example, assume that an interrupting device requests service. One of the interrupt request lines (IREQ0 - IREQ7) will be asserted and, as a result, the interrupt request is stored in the Interrupt Request Register (IRR). This interrupt is sent along to the CPU, provided that it is not masked out. The Group Interrupt line (GINT) is activated, notifying the CPU of the interrupt. The CPU must then distinguish which device needs servicing. The universality of the Am9519A becomes evident now. Most microprocessors begin an interrupt acknowledge cycle at this point that will prompt the interrupting device or controller to place a vector on the data bus. This vector is a specific op code that points the CPU to the interrupt handling software for the interrupting device. The vectors, of course, will vary from machine to machine. To account for these differences, the Interrupt Controller has a programmable Response Memory that can hold up to four bytes of information for each interrupt request line. During initialization, the Response Memory is loaded with the appropriate vectors, op codes, or data for the CPU in use. The Response Memory data is placed on the data bus during the interrupt sequence. The CPU reads this information, then processes the interrupt, and finally returns to normal processing activity.

Transferring Commands and Data

Data and Commands are exchanged between the CPU and the Interrupt Controller in a straightforward manner. A Chip Select line ($\overline{CS}$) is used to enable read and write operations. Chip Select is generated from the CPU address bus and, in this instance, is typically decoded from an I/O port address. $\overline{\text{CS}}$ conditions the use of the $\overline{\text{Read}}$ and $\overline{\text{Write}}$ signal lines (all three are negative level active). Data and Command transfers take place over the data bus under the control of these signals. Data and Commands are distinguished by the Control/$\overline{\text{Data}}$ (C/$\overline{\text{D}}$) line. A high level on this line will place write information into the Command Register; a read operation obtains information from the Status Register. If C/$\overline{\text{D}}$ is low, then read and write operations will involve registers previously selected by Command instructions. The C/$\overline{\text{D}}$ line is usually the low order bit of the Address Bus. This allows one port address to be used to represent data information and the next sequential port address to represent control information. The signal lines mentioned are usually exercised during the course of chip initialization and updating. When an interrupt first occurs, the Response Memory data is transferred on the data bus to the CPU. The Response Memory data will point the CPU to the software intended to process the current interrupt. Access to the Response Memory is controlled by the Interrupt Acknowledge ($\overline{\text{IACK}}$) line. Chip Select should be inactive during this time (100 nsec prior to the interrupt acknowledge cycle as recommended by the manufacturer).

Interrupt Controller Register Operation

Seven 8-bit registers, one 2-bit register, and a 32 × 8 Response Memory comprise the Universal Interrupt Controller storage facilities. These registers are discussed in detail below.

- *Command Register.* The 8-bit Command Register is used to store the most recent command sent to the Am9519A over the Data bus. The command that is stored will determine the operating mode of the Interrupt Controller. The commands are discussed in detail in the Commands section.
- *Mode Register.* The eight bits of the Mode Register are used to control some of the operational characteristics of the Interrupt Controller. A group of commands in the Controller Command Set are dedicated to loading these bits. In fact, there is no single command that will modify all the Mode Register bits at once. Figure II.7.3 shows the functional assignments of the Mode Register bits. Bits M0, M1, and M2 will be discussed later, where appropriate. Bits M3 and M4 are programmed at initialization to specify the active levels of the Group Interrupt (GINT) and

Figure II.7.3 Mode Register

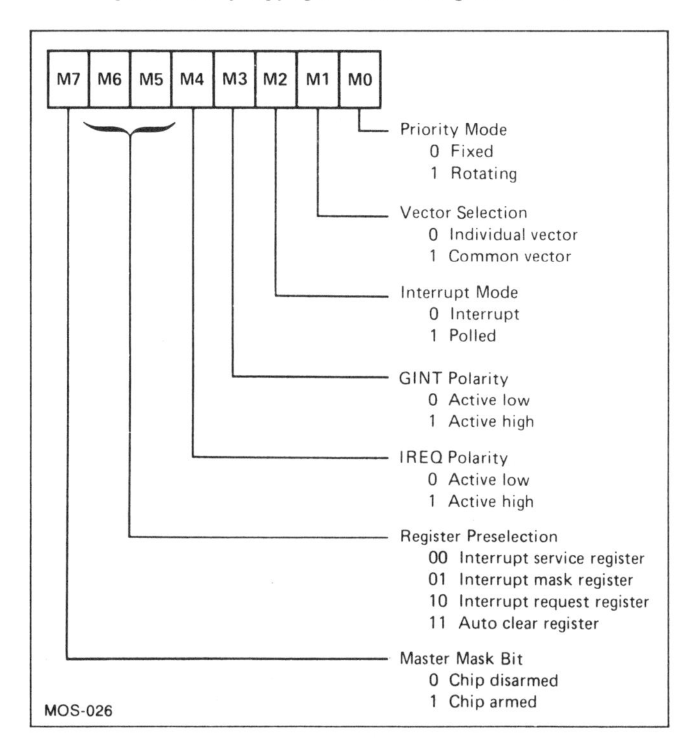

Interrupt Request (IREQ) lines. The ability to specify the levels of these signal lines allows an easy hardware interface to be created between the CPU, the Interrupt Controller, and the interrupting device. Bits M5 and M6 are grouped together to encode which one of four Interrupt Controlling registers will be read out to the data bus. The register selected will be read out to the data bus on the next Interrupt Controller read operation. The last bit, bit M7, is the Master Mask Bit. The effects of this bit, when set, are the masking out of all interrupt requests and the disabling of all lower priority chips. Although interrupts are masked out, they are still accepted on the IREQ lines and will be activated when the Master Mask bit is reset.

- *Interrupt Request Register (IRR).* There is one bit in the IRR for each of the eight Interrupt Request (IREQ) lines. A bit will be set when its associated IREQ line makes the transition to its preprogrammed active level. Under the proper conditions, a

Group Interrupt (GINT) will be generated subsequent to the setting of the IRR bit. The interrupting process, therefore, keys off of this register. The IRR bits can be cleared by several different methods: a software reset, execution of various Interrupt Controller commands, or automatically via an interrupt acknowledge from the processor. In addition, the IRR bits can be set by CPU software, allowing for software diagnostic testing of the interrupt system.

- *Interrupt Mask Register (IMR).* Setting a bit in the IMR will disable (mask) the individual interrupt associated with it. That is, an interrupt generated on an IREQ line and stored in the IRR will not activate the Group Interrupt line to the CPU if its corresponding bit in the Interrupt Mask Register is set. Individual interrupt levels can, therefore, be enabled or disabled under software control. If an interrupt is masked out by this process it can still be acted upon once the mask bit in the IMR is reset because the IRR will have stored the interrupting device's request for service. The bits in the IMR can be set, reset, or read to the data bus through various software commands either individually or on a byte basis. A chip reset sets all mask bits, thereby disabling all interrupts. To avoid potential problems, the manufacturer suggests disabling the CPU's interrupt system before modifying the Interrupt Mask Register. In this manner, an interrupt will not be in progress when the Mask Register or other interrupt hardware is being reprogrammed.

- *Interrupt Service Register (ISR).* For each bit in the Interrupt Request Register, there is a corresponding bit in the Interrupt Service Register that is used to handle the CPU interrupt response. The CPU will respond to a Group Interrupt signal by beginning an interrupt acknowledge cycle. During the course of this cycle, the CPU will assert the Interrupt Acknowledge ($\overline{\text{IACK}}$) input to the Interrupt Controller. At this point, priority-sensing logic within the Am9519A selects the interrupt with the highest priority for service. The interrupt selected may not be the interrupt that generated the initial Group Interrupt signal, since interrupts of higher priority may be generated in the time interval between GINT and $\overline{\text{IACK}}$. In any case, the bit in the IRR associated with the interrupt is cleared and the corresponding ISR bit is set with the arrival of $\overline{\text{IACK}}$. This frees up the IRR register for further interrupt requests while the currently running interrupt still receives service.

The ISR bit accomplishes another purpose. If the chip is configured for fixed priority mode (defined in the Mode Register), then all other interrupts of equal or lower priority are inhibited until the ISR bit is cleared. Only higher priority interrupts will generate GINT under these circumstances. Whether or not the CPU will respond to the new GINT signal depends upon the interrupt status of the CPU. If allowed, the higher priority interrupt will now succeed the original interrupt. If rotating priority is assigned, interrupts occur only on a rotating priority basis.

The bits in the ISR can be cleared by a command specific to the bit or through the Auto Clear feature. Clearing the bit with a command involves additional software in the interrupt servicing routine. The Interrupt Service Register first must be read to the data bus. This involves preselecting the ISR with the Mode Register, and a read ($C/\overline{D}=0$) operation. The ISR bits then need to be tested, and finally the appropriate command is issued to clear the bit. The Auto Clear feature will cause the ISR bit to clear before the end of the interrupt acknowledge cycle, eliminating the need for a software response to this condition.

- *Auto Clear Register (ACR).* The Auto Clear Register determines if the Auto Clear feature is in place as described above. Each ISR bit has an ACR bit. A one level in any ACR bit enables the Auto Clear feature for its corresponding ISR bit. A command preselects the Auto Clear Register for modification and, on the next write operation to the Interrupt Controller chip, parallel loads the ACR. The programmer can therefore select individual ACR bits, giving Auto Clear capability to selected interrupt levels. Auto Clear can also allow lower priority interrupts in a fixed priority system to assert GINT more readily.
- *Status Register.* The Status Register contains information on the internal state of the Am9519A. The information contained in the Status Register can be read by the CPU during a read operation when $C/\overline{D}$ is high. The eight bits of the Status Register depicted in Figure II.7.4 indicate the various states of the Am9519A. Bit S7 indicates the status of the Group Interrupt line. This signal is valid for both interrupt and polled modes of operation. When S7 indicates that at least one unmasked IRR bit is set, then S0, S1, and S2 can be checked for the highest priority pending interrupt. As an example, if S0, S1, S2 = 101, then IREQ5 is the pending interrupt.

Figure II.7.4 Status Register

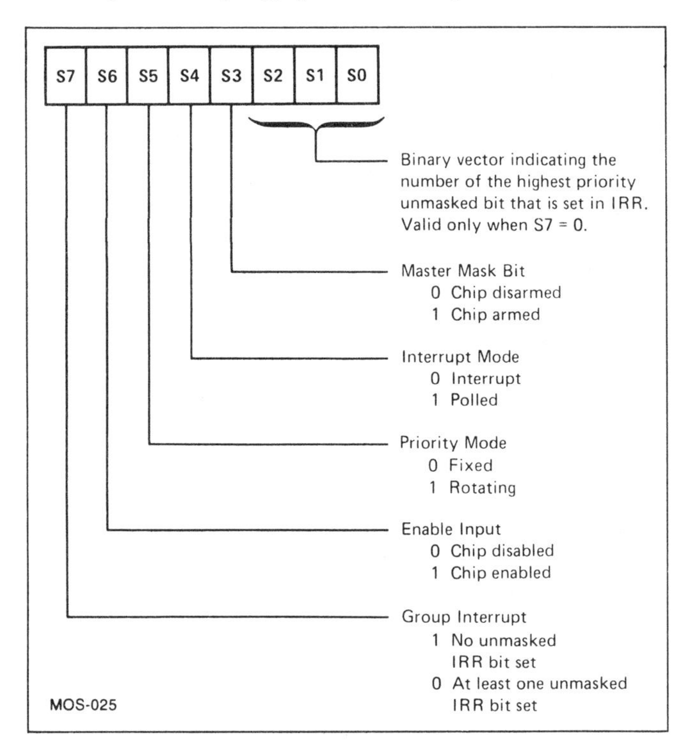

S3, S4, S5, and S6 provide information on modes of operation and enable/disable features as indicated in Figure II.7.4. The Status Register, at power up, will read 11000111. Software resets will not modify the Status Register.

- *Response Memory.* Each Interrupt Request line has at its disposal, a four byte segment of the 8 × 32 Response Memory (Fig. II.7.5). At initialization time, this RAM storage area is programmed with the vector, op code, instruction, or data that is appropriate to provide the interrupt response for the CPU in use. When an interrupt is acknowledged by the CPU, the segment of Response Memory corresponding to the interrupt generated will be read out to the data bus under control of IACK pulses. Depending on the method used to implement the interrupt, one to four bytes of information are needed by the CPU to process the interrupt. For example, in an 8085 system, a one-byte vector could be used to point to a predefined address in

Figure II.7.5 Response Memory

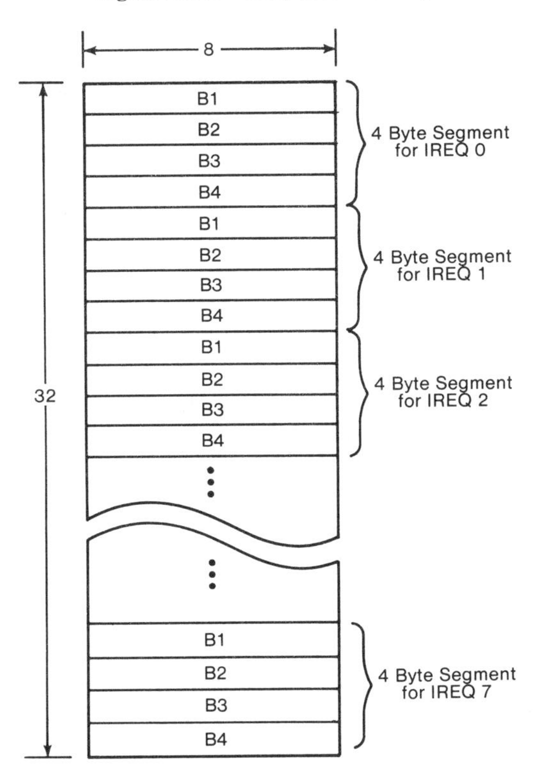

the CPU's memory space, which is where the interrupt software routine would begin. Alternatively, the 8085 could be presented with a three-byte CALL instruction from the Response Memory that directly points to an interrupt handling subroutine. With up to four bytes available for a response, a wide variety of processors can use the Universal Interrupt Controller.

Programming the Response Memory locations is accomplished through a command that specifies the number of bytes to be loaded as well as the interrupt level targeted for the bytes. After the command is issued, one to four bytes are loaded through subsequent write operations. During actual operation, response bytes are placed on the data bus in the same order as initially entered. All interrupt levels in the Response Memory should have at least one byte loaded, even if unused.

- *Byte Count Register.* The Byte Count Register is an 8 × 2 memory that stores the two byte length of the Response Memory data for each interrupt level. This register is loaded along with the Response Memory. During an interrupt, the byte counter will expect a number of $\overline{\text{IACK}}$ pulses equal to the count stored. This will allow the correct number of response bytes to be transferred to the CPU. The CPU will generate the $\overline{\text{IACK}}$ pulses as part of its software execution of the interrupt acknowledge cycle.

Interrupt Controller Commands

The Interrupt Controller Command Set gives the programmer the capability to set or clear the various bits in the registers (excluding Status) and Response Memory with great ease and flexibility. In addition, a software reset command is available. All commands are loaded by issuing a write operation with $C/\overline{D} = 1$. Figure II.7.6 summarizes the various Am9519A commands. Some of the commands have an X in certain bit positions. These are don't-care conditions, and the commands with don't-care bits initiate an explicit action within the Am9519A, such as CLEAR IMR or SET IRR. Other commands, such as CLEAR SINGLE IRR BIT, have bit positions labeled B0, B1, and B2. This three-bit field is used by the programmer to specify the actual bit in a register to be modified. For instance, setting B2, B1, B0 to 110 in the CLEAR SINGLE IRR BIT command would clear only bit six in the Interrupt Request Register.

Three of the commands are preselect commands. Their purpose is to point the next write operation data to a specific register in the Interrupt Controller. For example, issuing the PRESELECT ACR FOR WRITING command will set up the chip so that the data present in the next write operation will be loaded into the ACR. After this write operation, another command would typically be issued. Otherwise, succeeding write operations would reload the preselected register (ACR in this example). During the interval between the preselect command and the following write operation, read operations can be executed without problems.

The preselect command for the Response Memory has two fields for programmer use. The field labeled BY1 and BY0 is used to specify the byte count for the data to be loaded. Caution is needed here, since the binary code for this field does not translate to the decimal equivalent of the desired byte count. For example, coding BY1, BY0 = 00 indicates a byte count of one, whereas BY1, BY0 = 11 indicates a byte count of four. The field labeled L2, L1, L0 is used to represent the level of interrupt request. L2, L1, L0 = 011 indicates an interrupt level of three. In other words, the command would be preselecting the response

Figure II.7.6 Am9519A Command Summary

Command Code								Command Description	
7	6	5	4	3	2	1	0		
0	0	0	0	0	0	0	0	Reset (Clear IRR, ISR, ACR; set IMR)	
0	0	0	1	0	X	X	X	Clear all IRR and all IMR bits	
0	0	0	1	1	B2	B1	B0	Clear IRR and IMR bit specified by B2, B1, B0	
0	0	1	0	0	X	X	X	Clear all IMR bits (Enable)	IMR
0	0	1	0	1	B2	B1	B0	Clear IMR bit specified by B2, B1, B0	
0	0	1	1	0	X	X	X	Set all IMR bits (Disable)	
0	0	1	1	1	B2	B1	B0	Set IMR bit specified by B2, B1, B0	
0	1	0	0	0	X	X	X	Clear all IRR bits	IRR
0	1	0	0	1	B2	B1	B0	Clear IRR bit specified by B2, B1, B0	
0	1	0	1	0	X	X	X	Set all IRR bits	
0	1	0	1	1	B2	B1	B0	Set IRR bit specified by B2, B1, B0	
0	1	1	0	X	X	X	X	Clear highest priority ISR bit	ISR
0	1	1	1	0	X	X	X	Clear all ISR bits	
0	1	1	1	1	B2	B1	B0	Clear ISR bit specified by B2, B1, B0	
1	0	0	M4	M3	M2	M1	M0	Load Mode register bits 0-4 with specified pattern	
1	0	1	0	M6	M5	0	0	Load Mode register bits 5, 6 with specified pattern	
1	0	1	0	M6	M5	0	1	Load Mode register bits 5, 6 and set Mode bit 7	
1	0	1	0	M6	M5	1	0	Load Mode register bits 5, 6 and clear Mode bit 7	
1	0	1	1	X	X	X	X	Preselect IMR for subsequent loading from data bus	
1	1	0	0	X	X	X	X	Preselect ACR for subsequent loading from data bus	
1	1	1	BY1	BY0	L2	L1	L0	Load BY1, BY0 into byte count register and preselect response memory level specified by L2, L1, L0 for subsequent loading from data bus	

memory locations associated with an interrupt from input IREQ3. For write operations following this command, the write data would be loaded into the selected Response Memory locations equal to the byte count (i.e., if BY1, BY0 = 01, two bytes of data are loaded into the segment of Response Memory for IREQ3).

INTERRUPT VS. POLLED MODES OF OPERATION

The interrupt method described to this point is selected when Mode Register bit M2 = 0. Through this process, the CPU is notified of an interrupt condition by an active level on the GINT line. The CPU responds with $\overline{\text{IACK}}$ and the proper Response Memory locations are called upon to provide the correct interrupt vector. The vector (or instruction, data, etc.) is selected in the described fashion when another Mode Register bit is cleared. M1 = 0 specifies that the individual Response Memory locations are to be used when responding to an interrupt. With M1 = 1, a common vector mode is specified. Under this condition, all interrupts will receive the interrupt vector associated with IREQ0. This can be useful if all interrupting devices use the same interrupting software.

Referring back to Mode Register bit M2, it is possible to configure the Interrupt Controller chip for a polled method of interrupt handling by setting M2 to a high level. The CPU will no longer be notified of a pending interrupt through the GINT line because GINT will be forced to its inactive state. In fact, the interrupt process described will not be in effect at all. Instead, the CPU will have to poll (read) the Status Register for information on pending interrupts as well as handle all register clearing operations. In addition, in a cascaded system, a chip selected for polled operation is removed from the priority resolution circuitry. Priority will be in effect only for a single chip using polled interrupts. The polled method is more time consuming to the CPU, but does allow the programmer to use software generated interrupts. These can be particularly useful for testing the system interrupt structure.

SETTING INTERRUPT PRIORITY

Two priority methods are available in the Am9519A and are selected with Mode Register bit M0. M0 = 0 selects fixed priority mode, while M0 = 1 selects rotating priority mode. Using fixed priority, the IREQ inputs are assigned priority with IREQ0, the highest priority input, and

IREQ7, the lowest priority input. It is possible, with fixed priority, that lower priority inputs may never see a CPU response to their interrupt request because higher priority inputs constantly take precedence. Even if a low priority interrupt initiates the GINT line, the actual interrupt to receive service is the one with highest priority when the CPU returns the $\overline{\text{IACK}}$ signal. A higher priority interrupt could supersede the lower priority one during this interval. If this situation is intolerable, then rotating priority is preferable.

The relative priorities of the IREQ inputs are the same in rotating priority as they are in fixed priority. However, the interrupt in process will acquire lowest priority after service ends. Highest priority is now assigned to the IREQ input based upon its positional priority. For instance, assume that IREQ4 is being serviced. After service is complete, IREQ5 has highest priority while IREQ4 is now assigned lowest priority. With a rotating priority scheme, no IREQ input will have to wait longer than seven interrupt cycles before servicing.

SOME FINE POINTS

By cascading Interrupt Controller chips together, an interrupt structure with many interrupt levels can be obtained. The chips in the cascade need to be aware of each other so that priority resolution can take place throughout the whole structure. In a cascaded system, the Enable Out (EO) line of one chip is connected to the Enable In (EI) input of the next chip. During an interrupt acknowledge cycle, the internal circuitry of the Am9519A will, if appropriate, activate the EI/EO lines, selecting the highest priority chip in the cascade for service. This chip will then place its Response Memory information onto the data bus. A potential problem now exists. All other chips also see the $\overline{\text{IACK}}$ pulse and respond by placing their Response Memory information onto the data bus, corrupting the valid data. Another line, called Response In Progress ($\overline{\text{RIP}}$), solves this problem. $\overline{\text{RIP}}$ is a bidirectional, wired-OR (pull-up required) signal line. Each chip in the cascade is able to sense and control the $\overline{\text{RIP}}$ level. The chip that is responsible for the current interrupt in progress will pull $\overline{\text{RIP}}$ low. All other chips sense this level and are aware of a currently executing interrupt. They will therefore ignore the $\overline{\text{IACK}}$ signal that they receive, since it is not intended for them.

The Am9519A has an output, labeled Pause, that is used to extend the CPU's $\overline{\text{IACK}}$ pulse. Due to the timing delays possible from temperature, cascading chips, and resolving priority, the $\overline{\text{IACK}}$ pulse from the CPU may be too short to allow an interrupt response. $\overline{\text{Pause}}$ is essentially a mechanism by which wait states may be inserted into the CPU's processing cycle, thus extending the duration of the $\overline{\text{IACK}}$ pulse. $\overline{\text{Pause}}$ will remain active until the $\overline{\text{RIP}}$ pulse becomes active. Once $\overline{\text{RIP}}$ is asserted, the normal transfer of Response Memory information can take place.

$\overline{\text{Pause}}$, GINT, and $\overline{\text{RIP}}$ are all wired-OR circuits and require a pull-up resistor. The final interfacing of the Am9519A to a microprocessor chip will depend on the microprocessor chosen. In most cases, very little logic hardware is required between the Am9519A and the microprocessor for a functional system.

II.8

Programmable DMA Controller

PART NUMBER	Am9517A 8237A 8237A-4 8237A-5 82C37A-5
FUNCTION	PROGRAMMABLE DMA CONTROLLER
MANUFACTURERS	ADVANCED MICRO DEVICES INTEL
VOLTAGES	Vcc = +5
PWR. DISS.	1.5 W max
PACKAGE	40-pin DIP 40-pin LCC
TEMPERATURE	0°C → +70°C
FEATURES	FOUR INDEPENDENT DMA CHANNELS INDIVIDUALLY ENABLED AUTO INITIALIZATION, SOFTWARE DMA REQUESTS MEMORY-TO-MEMORY TRANSFERS END OF PROCESS INPUT DATA TRANSFER RATE UP TO 1.6 MBYTES/SEC
COMPATIBLE MICROPROCESSORS	8080 8085 8086 8088
FUNCTIONAL DESCRIPTION	THIS DMA CONTROLLER IS AVAILABLE IN VARYING PERFORMANCE GRADES. FOUR INDEPENDENT CHANNELS, UNDER CONTROL OF VARIOUS INTERNAL REGISTERS, ALLOW FOR THE HIGH SPEED EXCHANGE OF DATA BETWEEN EXTERNAL DEVICES AND CPU MEMORY. PRIORITY LOGIC ASSIGNS FIXED OR ROTATING PRIORITY AMONG THE DMA CHANNELS.

Figure II.8.1 8237A Pinout

Courtesy of Intel Corp.

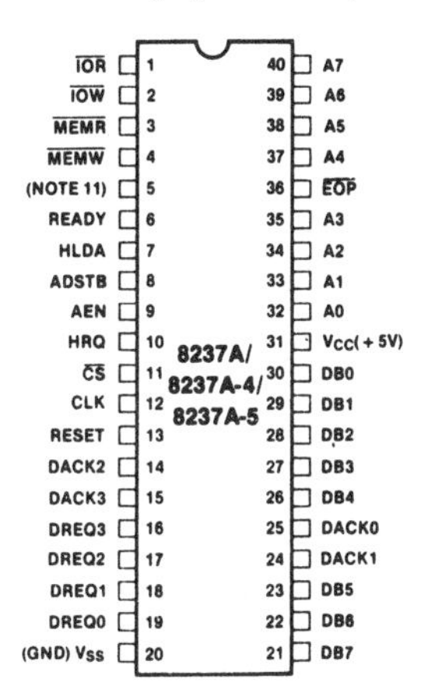

Figure II.8.2 8237A Block Diagram

Courtesy of Intel Corp.

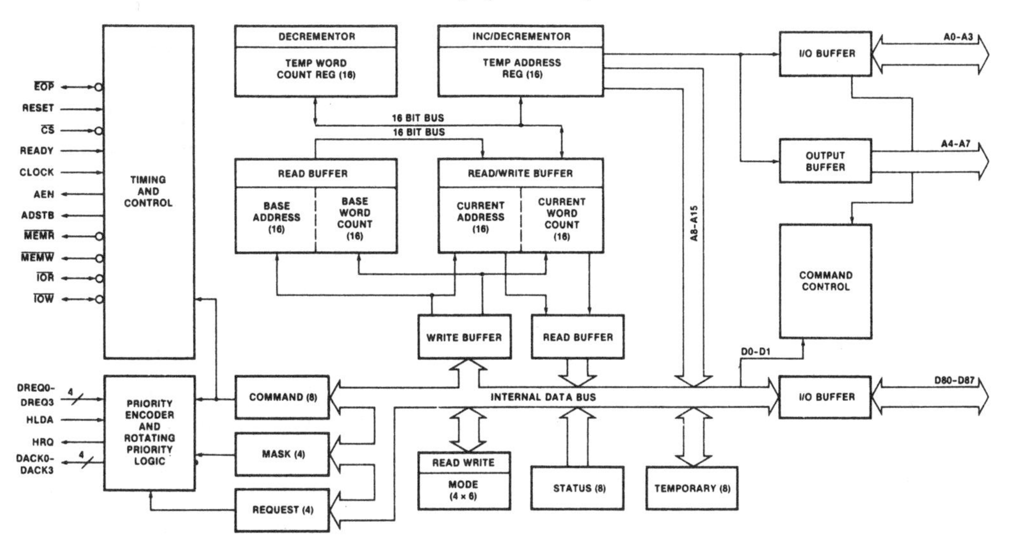

PIN NAME	PIN SYMBOL	PIN NUMBER	FUNCTION
CHIP SELECT	$\overline{\text{CS}}$	11	$\overline{\text{CS}}$, when low, enables the DMA Controller as an I/O device. Under these conditions, the CPU and controller can exchange programming information.
I/O READ	$\overline{\text{IOR}}$	1	During programming, $\overline{\text{IOR}}$ allows the CPU to read the DMA Controller register information. During DMA activity, this signal becomes a control output aiding the reception of data from an external device.
I/O WRITE	$\overline{\text{IOW}}$	2	During programming, $\overline{\text{IOW}}$ allows the CPU to write into the DMA Controller's registers. During DMA activity, this signal becomes a control output aiding the transferral of data to an external device.
CLOCK	CLK	12	This input is generally driven by the CPU clock at a rate between 3 MHz and 5 MHz. The actual rate depends on the performance range of the controller selected.
RESET	RESET	13	A high level on Reset will clear the Command, Status, Request, and Temporary registers and set the Mask register. The first/

PIN NAME	PIN SYMBOL	PIN NUMBER	FUNCTION
			last flip-flop is also cleared. Following the Reset, the controller is placed in the idle state.
DATA BUS 0	DB0	30	The data bus signal lines transfer commands and register information between the CPU and controller, contain the most significant byte of address for DMA transfers, and are the pathways for the data exchanged during a DMA cycle.
DATA BUS 1	DB1	29	
DATA BUS 2	DB2	28	
DATA BUS 3	DB3	27	
DATA BUS 4	DB4	26	
DATA BUS 5	DB5	23	
DATA BUS 6	DB6	22	
DATA BUS 7	DB7	21	
ADDRESS BUS 0	A0	32	These four address bits are bidirectional and, as outputs, are used to provide the lower four bits of the DMA address or, as inputs, to obtain register addresses for programming from the CPU.
ADDRESS BUS 1	A1	33	
ADDRESS BUS 2	A2	34	
ADDRESS BUS 3	A3	35	
ADDRESS BUS 4	A4	37	These four address bits are tri-state outputs providing the four most significant bits of the DMA address.
ADDRESS BUS 5	A5	38	
ADDRESS BUS 6	A6	39	
ADDRESS BUS 7	A7	40	
DMA REQUEST 0	DREQ0	19	External devices will assert their associated DREQ lines to request DMA service. The active levels on these lines are programmable. A Reset forces these lines to an active high level.
DMA REQUEST 1	DREQ1	18	
DMA REQUEST 2	DREQ2	17	
DMA REQUEST 3	DREQ3	16	

(*continued*)

PIN NAME	PIN SYMBOL	PIN NUMBER	FUNCTION
DMA ACKNOWLEDGE 0	DACK0	25	External devices are informed that a DMA cycle has been granted via their associated DACK line. The active levels on these lines are programmable. A reset forces the lines to active low.
DMA ACKNOWLEDGE 1	DACK1	24	
DMA ACKNOWLEDGE 2	DACK2	14	
DMA ACKNOWLEDGE 3	DACK3	15	
HOLD REQUEST	HRQ	10	Hold Request is generated by the DMA Controller in response to a valid DREQ signal. When validated, the CPU is placed into the Hold state, allowing DMA activity to begin.
HOLD ACKNOWLEDGE	HLDA	7	A high level on this input indicates that the CPU has entered the Hold state and that the system buses are now available for DMA.
ADDRESS ENABLE	AEN	9	This active high signal is used to enable an external latch containing the most significant portion of the DMA address. The line is also used to disable bus drivers during DMA activity.
ADDRESS STROBE	ADSTB	8	This active high signal is typically used to latch the transitory, most significant portion of the DMA address.
MEMORY READ	$\overline{\text{MEMR}}$	3	This tri-state line is asserted low by the control-

PIN NAME	PIN SYMBOL	PIN NUMBER	FUNCTION
			ler during a DMA system memory read.
MEMORY WRITE	$\overline{\text{MEMW}}$	4	This tri-state line is asserted low by the controller during a DMA write to system memory.
READY	READY	6	Ready is an active high input activated by slow memory or peripheral components to request additional cycle time. An active READY line will extend the $\overline{\text{MEMR}}$ and $\overline{\text{MEMW}}$ pulses.
END OF PROCESS	$\overline{\text{EOP}}$	36	As an output, $\overline{\text{EOP}}$ signals the end of DMA service. As an input, an active low signal forces DMA service to terminate. $\overline{\text{EOP}}$ should be tied high with a pull-up resistor if not used.
Vcc	Vcc	31	+5 V.
Vss	Ground	20	Ground.
Pin 5	*	5	This line should always be at a high logic level. A pull-up resistor can be used, although an internal pull-up will hold the level if the input is left floating.

Direct Memory Access (DMA) can increase computer system data throughput between memory and external devices. A DMA controller will not only simplify the hardware interface involved in the implementation of such a system, but it will also add a level of versatility, through programming, that will ease the software burden involved.

UNDERSTANDING DMA CONTROLLER OPERATION

This DMA Controller can be viewed as having three operational states: inactive, programming, and DMA mode.

When the controller is in the inactive state, the chip will not interfere with any processor activity. In this case, the chip has tri-stated all lines connected to the system buses and has all control line levels in the inactive state. During this mode of operation, the chip will periodically check for an active level on its Chip Select ($\overline{CS}$) or DMA request (DREQ) lines. An active level on one of these lines will place the chip into another operational category.

A low level on Chip Select will put the DMA Controller into a programmable condition. The internal registers can now be read or written, allowing the chip to be configured for any desired method of operation. The registers involved can be viewed as sixteen distinct I/O ports accessible through the processor's I/O command structure. Simple CPU output instructions can therefore be utilized to load the internal registers of the DMA Controller. The Hold Acknowledge line (HLDA) must be inactive at this time.

DMA mode, or the active cycle, occurs when an external device requests DMA service. The device will request service by asserting its DMA Request (DREQ) line. The chip responds by asserting its Hold Request (HRQ) line to the CPU, assuming that the DMA channel requesting service has been enabled. Once the CPU has acknowledged the Hold Request signal (via HLDA), the DMA transfer begins. The CPU will relinquish the system buses and deactivate its control signals while the DMA Controller takes over. The controller provides all the necessary addresses and signals as it carries out the preprogrammed DMA procedure.

How Controller Registers Guide DMA Operation

The DMA Controller has four independent DMA channels. Each channel is attached to one external device and is programmed to handle the DMA requirements of that device. A set of control registers is provided for each channel for this function. Figure II.8.1 shows the pinout diagram of the DMA Controller while Figure II.8.2 shows the internal block diagram.

Figure II.8.3 Request Register
Courtesy of Intel Corp.

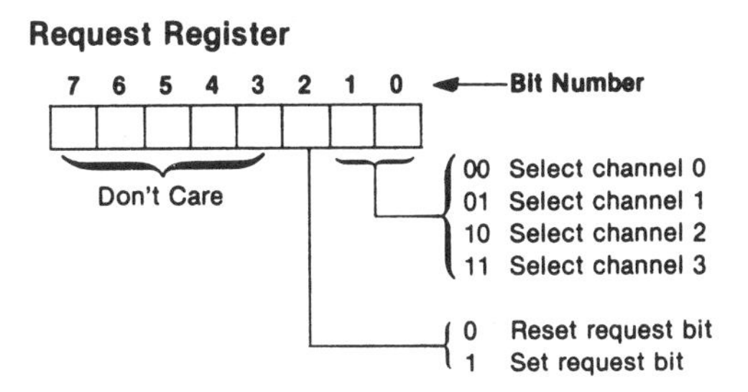

A typical DMA transfer involves moving data bytes from an external device into system memory. Using this premise, the register activity involved can be investigated.

Each channel has a bit in the Request Register designed to store the fact that a DMA request has occurred. The request can come from the DMA request lines (DREQ0-DREQ3), or directly from the CPU via software. The Request Register, shown in Figure II.8.3, can be set or cleared using software to control bits 0, 1, and 2. The two encoded low order bits are set to a value that selects the desired channel. Bit 2 then determines if the selected Request bit is set or reset. Setting the bit initiates a software DMA request. Software requests are non-maskable; hardware requests through the DREQ lines may be masked using the Mask Register. The Request Register is also cleared by a Reset. Individual bits are cleared when an external End Of Process ($\overline{\text{EOP}}$) is received or when the DMA procedure ends.

Each DMA channel can be disabled with mask bits contained in the Mask Register. Each channel has its own mask bit that is set or cleared by a software command. Setting a mask bit disables its associated DMA channel from operation. The mask bits are also set by a Reset or after a DMA procedure, provided that the Autoinitialize feature is not invoked. The Mask Register bits can be set individually, in a fashion similar to the Request Register, as shown in Figure II.8.4a. The two low order bits select the desired channel while bit 2 determines the level. Another format, shown in Figure II.8.4b, allows all bits to be modified in a single software command (Write All Mask Register). Each of the four least significant bits is allocated to a channel and may be set or reset as indicated.

The Command Register dictates the overall operation of the DMA Controller (Fig. II.8.5). Each bit in this 8-bit register is dedicated to a single function. Bit 2, for instance, can enable or disable controller

Figure II.8.4 Mask Register Formats

Courtesy of Intel Corp.

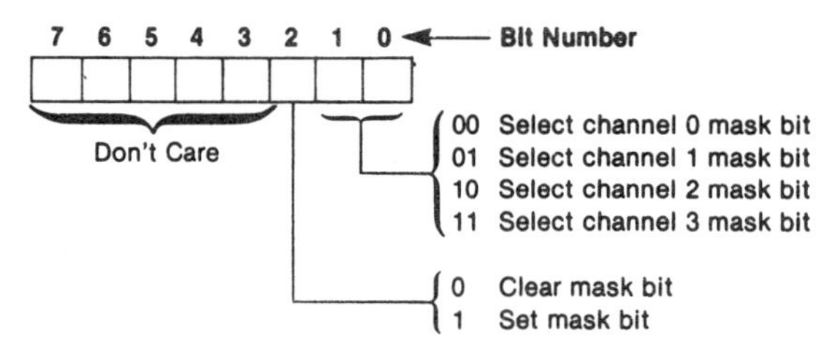

(a) Individual Mask Selection

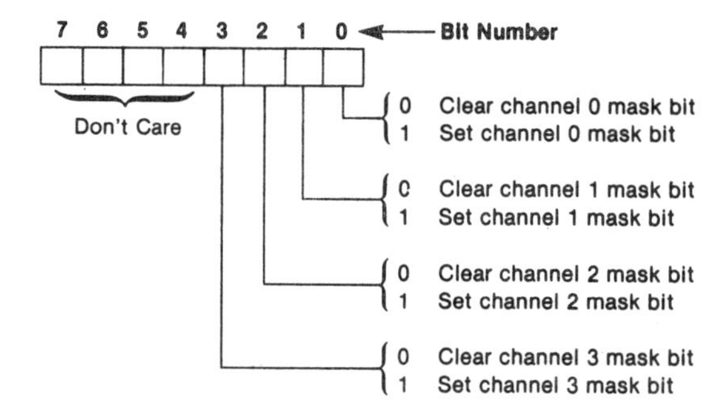

(b) Mask Bits Set with a Single Command

operation. When disabled, the controller can be programmed, but DMA activity is inhibited. Bit 4 selects the priority scheme desired. Fixed priority assigns the greatest priority to channel 0 and the least priority to channel 3. Using this setup it is possible that channel 3 could be preempted from DMA service by the other higher priority channels. Therefore, the priority selection technique should be considered carefully. The alternate priority selection scheme is one of rotating priority. The initial channel priority assignment is the same as the fixed priority assignment; that is, channel 0 has the highest priority and channel 3 the lowest. After activity begins, the most recently serviced channel is assigned lowest priority and will have to wait for all other channels to be serviced before receiving service again. Of course, if the other channels are not active, the most recently serviced channel, even with lowest priority, could receive service from the controller. Bits 6 and 7 establish the active levels on both the DMA Request (DREQ) lines and the DMA Acknowledge (DACK) lines. This ability to select the active levels of key signal lines will simplify the hardware interface. The other Command Register bits are discussed in later sections of the text.

Figure II.8.5 Command Register
Courtesy of Intel Corp.

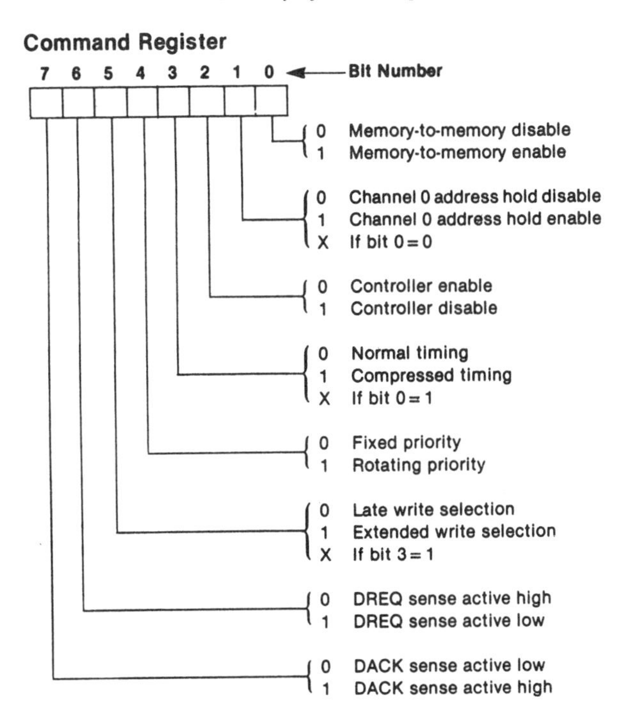

A Mode Register is assigned to each channel which configures the channel for individual operating characteristics. The format for programming this register is given in Figure II.8.6. Low order bits 0 and 1 select the desired channel for programming. Bits 2, 3, 6, and 7, control the data transfer method and are discussed later. Bit 4 will select if Autoinitialization is to be used. Autoinitialization can save programming overhead, as well as time, by allowing the DMA channel to automatically reconfigure itself back to its original starting state after DMA operation. Bit 5 gives the programmer control over the DMA address. The address can be made to automatically increment or decrement after each byte of data is transferred (see example that follows).

As an example of DMA address control, assume that an upcoming DMA transfer requires that data be moved from an external device to a block of system memory. Since the DMA chip rather than the CPU controls the system buses in this example, the controller must provide the memory addresses where the data is to be stored. Two registers per DMA channel are set aside for this purpose. The Current Address Register holds the 16-bit value of the address used for a DMA transfer. By

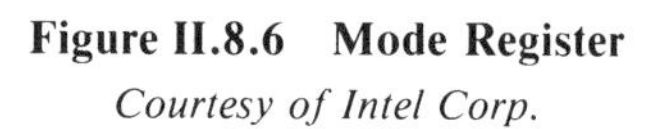

Figure II.8.6 Mode Register
Courtesy of Intel Corp.

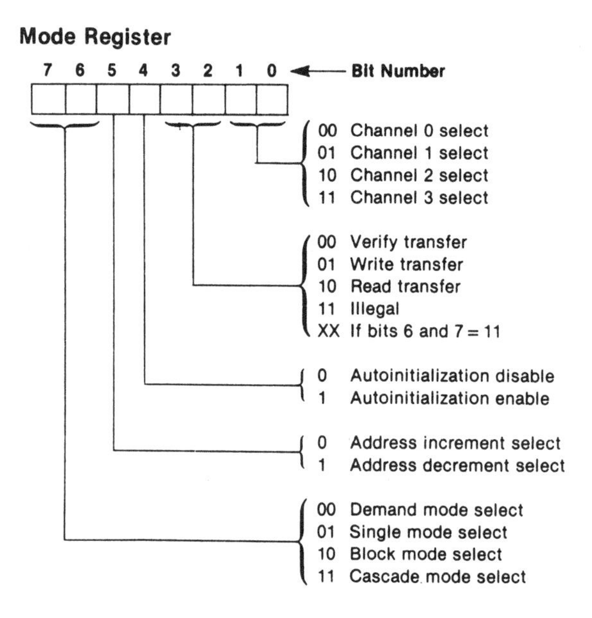

programming with the Mode Register the current address register can be made to automatically increment, decrement, or reestablish its original value. The register can also be read or written from the CPU. Since this register is 16 bits wide, two successive read or writes are necessary to program or access its contents. During actual DMA usage, the whole 16-bit entity is output on the address bus. A second register, called the Base Address Count Register, stores the initial value of address that was programmed into the Current Address Register. There is a Base Address Count Register for each Current Address Register. During a sequence known as Autoinitialize, the value in the Base Address Count Register is placed back into its corresponding Current Address Register, reestablishing the original DMA address. On the next DMA request to this channel, the proper memory address will already be in place. The Base Address Count Register is programmed in parallel with the Current Address Register and cannot be read by the microprocessor.

The number of data bytes that are transferred during a DMA operation must also be specified. This quantity is controlled by the 16-bit, Current Word Register. This register holds the value that is one less than the number of bytes to be transferred. For example, if the register is programmed with the number 16, then 17 bytes of data will actually be transferred. This is due to the fact that internally, the final Current Word Register value is detected as the register cycles from all zeros to all ones. This is referred to as reaching the Terminal Count (TC). When the Terminal Count is reached, all DMA data bytes have been transferred. As with the addressing registers, there is a companion to the Current Word Register called the Base Word Count Register. This register will hold the original value of the word count and is used in an Autoinitialize sequence to reestablish the value in the Current Word Register. The Base Word Count Register cannot be read by the CPU and, like its addressing counterpart, is written along with the Current Word Register.

Sometimes the DMA transfer mode selected involves the transfer of data from one memory location to another. This is called a memory-to-memory transfer and will necessitate temporary, on-chip storage for the data (see the heading, "Memory-to-Memory Transfers"). Since only one memory address can be accessed at a time, a register, appropriately called the Temporary Register, is provided for the temporary storage of the data. When data is read from a memory location under DMA control, it is first stored in the Temporary Register. After the data destination address is provided, the data is taken from the Temporary Register and moved into its new location. This register will always hold the last byte of data transferred in a memory-to-memory transfer unless a Reset occurs, in which case the register is cleared. The CPU can also read this register.

The Status Register, shown in Figure II.8.7, indicates all pending DMA requests as well as all channels that have reached a Terminal Count. The Terminal Count indicated may have occurred normally or by an external EOP signal. The Status Register is cleared by a Reset or a Status read operation.

Figure II.8.7 Status Register

Courtesy of Intel Corp.

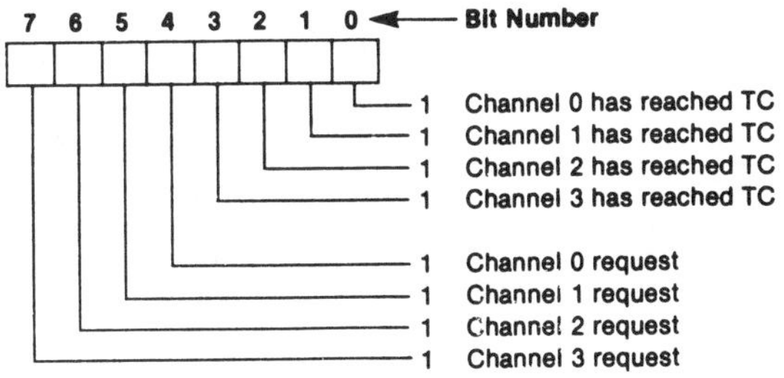

7 6 5 4 3 2 1 0 ← Bit Number

1 Channel 0 has reached TC
1 Channel 1 has reached TC
1 Channel 2 has reached TC
1 Channel 3 has reached TC

1 Channel 0 request
1 Channel 1 request
1 Channel 2 request
1 Channel 3 request

PROGRAMMING WITH DMA CONTROLLER COMMANDS

The bulk of the programming for this chip involves reading from or writing to the various registers. The registers are treated as I/O ports and can therefore be accessed using the CPU's I/O instructions such as INPUT and OUTPUT. For instance, to load the Command Register with a bit pattern, the programmer first places the appropriate bit pattern into the CPU's accumulator and then issues the OUTPUT instruction. Each I/O instruction needs a particular device code associated with it to specify a unique I/O port. Figures II.8.8a and II.8.8b show the assignment of port addresses used with the DMA Controller for the purpose of distinguishing between the registers. Note that address bits A0, A1, A2, and A3 actually indicate the register that will be read or written. Address bits A4 through A7 to comprise the Chip Select circuitry. Using the Command Register as an example, the OUTPUT instruction used would have to specify a device code of XXXX1000.

There are three software instructions that do not need any CPU data to be programmed. In these three, the port address alone is sufficient to enact the proper software response. The Master Clear command, as an example, merely needs the instruction OUT XXXX1101, (where X depends on the address space assigned the chip), to execute this command. The three commands that function in this way are:

- *Clear First/Last Flip-Flop.* This flip-flop is used in conjunction with the Address and Word Count Registers. Since these registers are 16-bit registers, accessed 8-bits at a time, the flip-flop is used to insure that the low order byte and then the high order register bytes are loaded in sequence. Clearing the flip-flop allows the low order byte of the Address or Word Count Register to be loaded with the first OUTPUT instruction issued. All subsequent access to these registers will then proceed in the proper sequence. The flip-flop can also be cleared using the Master Clear command discussed next.
- *Master Clear.* This is the software version of a hardware Reset. This instruction places the chip into an inactive state and clears the Command, Status, Request, and Temporary registers, the First/Last Flip-Flop, and sets all bits in the Mask Register.
- *Clear Mask Register.* This command clears all mask bits, enabling the DMA channels.

Figure II.8.8 Port Address Assignments

Courtesy of Intel Corp.

Signals						Operation
A3	A2	A1	A0	$\overline{IOR}$	$\overline{IOW}$	
1	0	0	0	0	1	Read Status Register
1	0	0	0	1	0	Write Command Register
1	0	0	1	0	1	Illegal
1	0	0	1	1	0	Write Request Register
1	0	1	0	0	1	Illegal
1	0	1	0	1	0	Write Single Mask Register Bit
1	0	1	1	0	1	Illegal
1	0	1	1	1	0	Write Mode Register
1	1	0	0	0	1	Illegal
1	1	0	0	1	0	Clear Byte Pointer Flip/Flop
1	1	0	1	0	1	Read Temporary Register
1	1	0	1	1	0	Master Clear
1	1	1	0	0	1	Illegal
1	1	1	0	1	0	Clear Mask Register
1	1	1	1	0	1	Illegal
1	1	1	1	1	0	Write All Mask Register Bits

(a) By Operation

Register	Operation	Signals						
		$\overline{CS}$	$\overline{IOR}$	$\overline{IOW}$	A3	A2	A1	A0
Command	Write	0	1	0	1	0	0	0
Mode	Write	0	1	0	1	0	1	1
Request	Write	0	1	0	1	0	0	1
Mask	Set/Reset	0	1	0	1	0	1	0
Mask	Write	0	1	0	1	1	1	1
Temporary	Read	0	0	1	1	1	0	1
Status	Read	0	0	1	1	0	0	0

(b) By Register

DMA TRANSFER MODES

The DMA Controller can be programmed for several types of DMA transfers. The DMA transfer will begin when the requesting device asserts its DREQ line. The controller then raises the Hold Request (HRQ) line to the CPU and, eventually, receives Hold Acknowledge (HLDA) signifying that the CPU has freed itself from the system buses. The DMA Controller then informs the requesting device that DMA may begin by activating DACK (DMA Acknowledge). To ensure a proper response, the requesting device should keep DREQ active until it receives the DACK signal. The actual transfer that takes place depends upon the Mode Register (bits 6 and 7) programming that governs the channel in use. (Fig. II.8.6).

Single Transfer Mode allows one byte of data to be transferred. DREQ must be reactivated after the data transfer in order for another transfer to begin. In this mode of operation, keeping DREQ constantly active still only results in a single data transfer, since the controller will release HRQ after the first transfer is complete. The Current Address Register is decremented or incremented, according to programming, by one after the transfer. The Current Word Register is decremented, also by one, and will provide a Terminal Count (TC) signal when the decrement causes the count to change from all zeros to all ones. Autoinitialize, if envoked, restores the original address and word count values when TC is reached.

Data may be continuously transferred in Block Transfer Mode. The transfer continues until TC is reached or an external End of Process ($\overline{\text{EOP}}$) signal is received. Autoinitialize can be programmed to be operative in this mode. DREQ must remain active at least until DACK is asserted.

Using the Demand Transfer Mode, data is continually transferred if DREQ is held active. If DREQ is made inactive, or if TC is reached, or if an external $\overline{\text{EOP}}$ is received, then the transfer is terminated. If the sending device runs out of data, it can reestablish the DMA link by reasserting DREQ when more data becomes available. Autoinitialize occurs with a TC or $\overline{\text{EOP}}$.

The number of DMA channels can be increased using Cascade Mode. Figure II.8.9 illustrates this. One DMA Controller is used to create a priority network around which all others will interface. This is referred to as the first level device. This chip is connected to the CPU (through the HRQ and HLDA lines) and to the other DMA Controllers. The DMA Controllers are connected using HRQ to DREQ, and DACK to HLDA. The first level device has only three functions: to collect DMA requests, establish priority, and request the Hold state from the CPU. The other controllers, attached to the I/O devices requesting service, will provide the addresses and control signals used during the actual DMA transfer.

HOW CONTROL SIGNALS ARE ACTIVATED FOR DMA

An 8-bit latch is used externally with the DMA Controller to accommodate the multiplexed nature of the data bus. Addressing pins are available on the chip for the lower byte but, to conserve pins, the high

Figure II.8.9 Cascading DMA Controllers
Courtesy of Intel Corp.

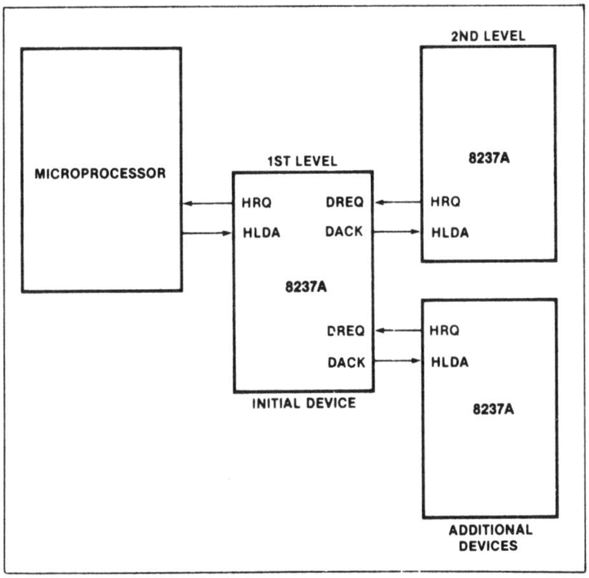

order address byte is multiplexed on the data bus. During the course of a DMA cycle, the high order portion of the address is present on the data bus for only a short time. The external latch is used to capture this information for use throughout the cycle. The Address Strobe (ADSTB) control line is active when the high order byte is present on the data bus and is used as a latching signal for this information. Another signal, Address Enable (AEN), is also active at this time and is used as a tri-state enable.

When a DMA transfer entails the movement of data from an I/O device to memory or from memory to an I/O device, an unusual combination of control signals will be active. In the case of a transfer from I/O to memory, called a Write Transfer, the DMA Controller will activate both $\overline{\text{IOR}}$ (I/O Read) and $\overline{\text{MEMW}}$ (Memory Write) at the same time. This occurrence never happens when the CPU is in control of these lines but makes sense in the case of a DMA transfer. The I/O device is designed to present data to the data bus via $\overline{\text{IOR}}$ and the memory is structured to accept the data via the $\overline{\text{MEMW}}$ signal. The data is not stored nor transferred through the controller itself for this kind of a transfer. The controller merely guides the flow of data with the appropriate signals and addresses. A similar situation occurs for data moving from memory to the I/O device. This is called a Read Transfer.

In this instance, $\overline{IOW}$ (I/O Write) and $\overline{MEMR}$ (Memory Read) are asserted at the same time. It is possible to generate these control signals, along with necessary addresses, without actually transferring any data. This is referred to as the Verify transfer type and allows the system to be tested. The response to signals such as TC and $\overline{EOP}$ can be evaluated using this type of transfer. The Read, Write, and Verify transfer types can be programmed using bits 2 and 3 of the Mode Register.

Transfer rates can be increased using Compressed Timing. This option eliminates one of the clock periods used during the DMA cycle, resulting in a shorter cycle time. Compressed Timing can be used with fast access time memory chips. Bit 3 of the Command Register selects normal or compressed timing.

Some data transfers, such as those using Block and Demand Transfer modes, will require sequential memory addresses. Recall that the high order portion of the address is stored in external latches. This part of the address changes once for every 256 times that the low order byte changes. Since time is required in the DMA cycle to update the high order byte, time is wasted if the high order byte doesn't change. The DMA Controller automatically supresses the DMA cycle state used for this update if it is not needed.

Some microprocessor components are not fast and the DMA Controller can also respond properly to them. If a slow device cannot complete its function during the time of a normal DMA cycle, it can request wait states from the controller to gain additional cycle time. The slow device raises the Ready line to accomplish this. Write times may also be modified using bit 5 (Late Write/Extended Write) of the Command Register.

Memory-to-Memory Transfers

While most DMA transfers involve the movement of data between a high speed I/O device and memory, it can be beneficial having the capability to move data from one set of memory addresses to another. This type of memory-to-memory transfer is supported by this DMA Controller.

A memory-to-memory transfer takes twice the number of cycles as other DMA transfers. In addition, the controller is used as a temporary storage facility for the data. Channel zero is programmed to be the source channel for the data and channel one is programmed to be the destination channel (bit 0 of the Command Register enables this capability).

A DREQ on channel zero starts the procedure. When HLDA is

received, the Current Address Register for channel zero provides the address for the source data. The data is stored temporarily in the DMA Controller's Temporary Register since the memory can only read or write at a given time. The $\overline{\text{MEMR}}$ line is activated to read the data. Once stored in the Temporary Register, $\overline{\text{MEMW}}$ is activated to write the data back to memory. The Current Address Register for channel one provides the destination address. Using bit 1 of the Command Register, the address for channel zero can be made to remain the same, allowing the same byte of data to be stored in all the addresses specified by channel one. This would be useful for a memory-fill operation of constant data.

II.9

Floppy Disk Controller

PART NUMBER	FDC765A uPD765 765A 8272A
FUNCTION	FLOPPY DISK CONTROLLER
MANUFACTURERS	STANDARD MICROSYSTEMS NEC ROCKWELL INTEL

VOLTAGES	Vcc = +5	**PWR. DISS.**	600 mW
PACKAGE	40-pin DIP	**TEMPERATURE**	0°C → +70°C

FEATURES

- SINGLE AND DOUBLE DENSITY RECORDING FORMATS
- PROGRAMMABLE SECTOR LENGTH
- CONTROLS UP TO FOUR FLOPPY OR MINI-FLOPPY DRIVES
- DMA OR NON-DMA TRANSFER MODES

COMPATIBLE MICROPROCESSORS MOST MICROPROCESSORS

FUNCTIONAL DESCRIPTION

THIS FLOPPY DISK CONTROLLER, COMBINED WITH AN EXTERNAL PHASE LOCKED LOOP AND WRITE PRECOMPENSATION CIRCUITS, PROVIDES ALL THE CONTROL FUNCTIONS TO INTERFACE UP TO FOUR FLOPPY DISK DRIVES WITH A MICROPROCESSOR SYSTEM. THE CONTROLLER IS COMPATIBLE WITH BOTH THE IBM 3740 SINGLE DENSITY (FM) AND THE IBM SYSTEM 34 DOUBLE DENSITY (MFM) RECORDING FORMATS.

Figure II.9.1 8272A Pinout

Courtesy of Intel Corp.

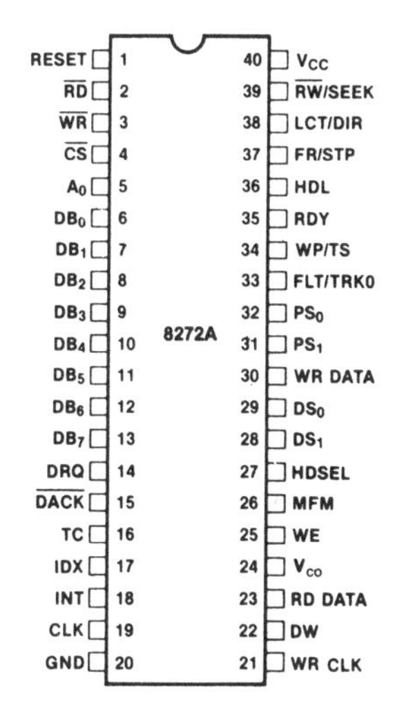

Figure II.9.2 8272A Block Diagram

Courtesy of Intel Corp.

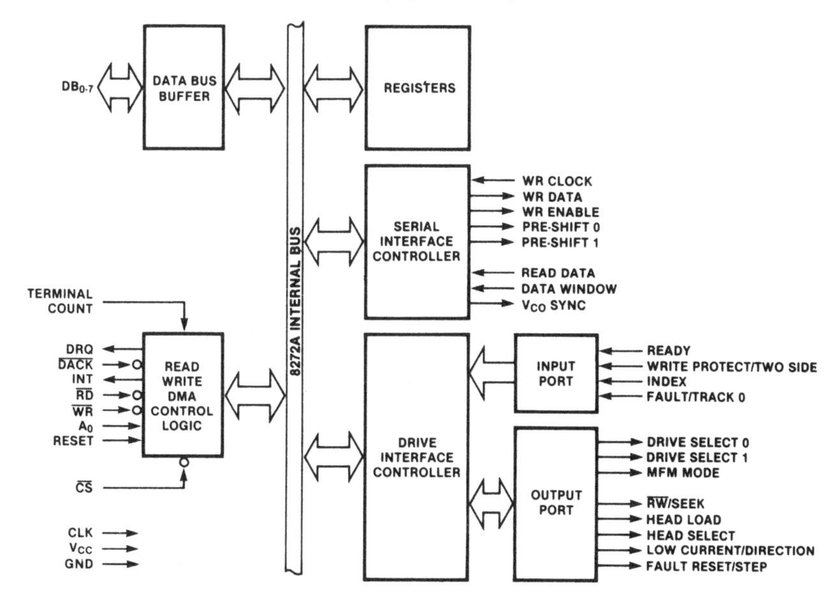

PIN NAME	PIN SYMBOL	PIN NUMBER	FUNCTION
CHIP SELECT	$\overline{CS}$	4	Chip Select, when low, enables the FDC (Floppy Disk Controller) for both Read and Write operations with the CPU.
READ WRITE	$\overline{RD}$ $\overline{WR}$	2 3	When low (only one line can be low at a time), these two CPU control lines enable the FDC to receive data from or send data to the CPU. This exchange of data takes place over the data bus.
DATA/STATUS	A0	5	This select line, typically controlled by an address bit, selects the main Status Register for a Read operation when low or selects the Data Register for a Read or Write operation when high.
RESET	RESET	1	A high level on this line places the FDC into the idle state and forces the Floppy Disk Drive output lines low.
DATA BUS 0 DATA BUS 1 DATA BUS 2 DATA BUS 3 DATA BUS 4 DATA BUS 5 DATA BUS 6 DATA BUS 7	DB0 DB1 DB2 DB3 DB4 DB5 DB6 DB7	6 7 8 9 10 11 12 13	All programming commands and data are exchanged between the FDC and the CPU over these lines.

PIN NAME	PIN SYMBOL	PIN NUMBER	FUNCTION
DMA REQUEST	DRQ	14	This line is raised high by the FDC to signal the CPU (via a DMA Controller) that a DMA transfer is desired.
DMA ACKNOWLEDGE	$\overline{\text{DACK}}$	15	This line is brought low by a DMA Controller, indicating the start of a DMA cycle.
TERMINAL COUNT	TC	16	A high level on this line indicates that the DMA Controller has terminated the DMA transfer.
INTERRUPT	INT	18	The interrupt line is raised high by the FDC to signal an interrupt request.
CLOCK	CLK	19	An 8-MHz clock is used for floppy disks. A 4-MHz clock is used for mini-floppies.
INDEX	IDX	17	This line from the FDD (Floppy Disk Drive), when high, indicates the beginning of a disk track.
READY	RDY	35	The FDD indicates that it is ready for data transfer when this line is high. This line is not used with mini-floppies and should be tied high.
WRITE PROTECT/ TWO-SIDE	WP/TS	34	Depending on the mode of operation, the FDD signals

(*continued*)

PIN NAME	PIN SYMBOL	PIN NUMBER	FUNCTION
			the FDC of Write Protection or Two-Sided Media.
FAULT/TRACK 0	FLT/ TRK0	33	Depending on the mode of operation, the FDD signals the FDC of a fault or that the RD/WR head is over Track 0 with this line.
DRIVE SELECT 0 DRIVE SELECT 1	DS0 DS1	29 28	The encoded information on these lines is used to select one of the four disk drives.
MFM MODE	MFM	26	A high level on this pin places the FDC into MFM (Double Density Format) while a low level configures the FDC for FM (Single Density Format).
READ WRITE/SEEK	$\overline{RW}$/ SEEK	39	A zero on this output line signals the FDD that a Read or Write operation is desired; a one signals a Seek operation.
HEAD LOAD	HDL	36	A high level on this line signals the FDD to perform a Head Load.
HEAD SELECT	HDSEL	27	Used to select the Read/Write head in a FDD with double-sided media. A low selects head zero; a one selects head one.
LOW CURRENT/ DIRECTION	LCT/ DIR	38	This line is used to reduce the write current level when writing to the inner

PIN NAME	PIN SYMBOL	PIN NUMBER	FUNCTION
			tracks of the floppy disk or to determine the direction of head travel in Seek mode.
FAULT RESET/STEP	FR/STP	37	In Read/Write mode this output, when active (high), will clear the Fault flip-flop. In Seek mode, the step pulses used to move the head are obtained from this line.
WRITE CLOCK	WR CLK	21	This input clock is 500 kHz for FM format and 1 MHz for MFM format. This clock is necessary for both Read and Write modes and requires a pulse width of 250 nsec.
WRITE DATA	WR DATA	30	Write Data is the serial data line to the FDD.
WRITE ENABLE	WE	25	WE enables the FDD for Write operations.
PRECOMPENSATION 0	PS0	32	These two lines determine the status of Write precompensation by specifying early, late, and normal Write times.
PRECOMPENSATION 1	PS1	31	
READ DATA	RD DATA	23	Read Data is the serial data line from the FDD.
DATA WINDOW	DW	22	This input line comes from an external Phased Locked Loop (PLL) and is used as

(*continued*)

PIN NAME	PIN SYMBOL	PIN NUMBER	FUNCTION
			a timing signal to sample data from the FDD.
VCO SYNC	VCO	24	A high level on this output enables the PLL VCO (Voltage-Controlled Oscillator).
DC POWER	V_{CC}	40	+5 V DC.
GROUND	GND	20	Ground.

AN INTRODUCTION TO FLOPPY DISK TECHNOLOGY

Floppy disks are used in micro-, mini-, and mainframe-computer systems to provide low cost, high density storage. Often referred to as removable magnetic media, a floppy disk also furnishes the means for distributing application, microcode, and operating system software to the users' computer systems.

Most people are familiar with the 8-, 5.25-, and 3.5-inch floppy disk sizes currently in use. The 5.25-inch disk is called a mini-floppy, while the 3.5-inch disk is called a micro-floppy. Enclosed within the protective jacket of each of these devices is a thin, plastic disk coated with a magnetic substance. Information is recorded onto and retrieved from this magnetic surface. The storage area on the disk is organized into many concentric circles on either one or both sides of the disk. These circles are called tracks. Data stored onto these tracks varies in organization from computer to computer, but, in general, tracks are usually broken down into sectors to form the basic element of storage. Tracks/sectors are therefore formatted before use to conform with the peculiarities of the computer system. When placed into a disk drive, the plastic disk is made to spin inside its jacket at a fixed rate of speed (360 rpm is typical). A read/write head, similar in function to those used in cassette recorders but optimized for pulse recording, makes contact with the spinning disk through an access hole in the protective jacket. This head assembly can be made to move toward the center of the disk or back toward the outer edge of the disk as it seeks the desired track for its next read or write operation. The head is a part of the floppy disk drive (FDD). The floppy disk drive contains the motors and mechanical contrivances which accurately position the read/write heads. The movement of data between the computer and the FDD takes place through an interface that converts binary signals to the proper form for magnetic recording and vice versa. It also keeps track of the head position as well as performing various other formatting/calibration duties. This interface is highly complex in nature. A floppy disk controller (FDC), such as the FDC765A, uPD765, 765A, or 8272A, is a single chip solution to the logical requirements of this interface.

Why a Floppy Disk System Uses Additional Components

As is the case with most complex peripheral chips, the floppy disk controller (FDC), illustrated in Figures II.9.1 and II.9.2, requires programming to initialize the chip for specific operation. Programming FDC registers is accomplished as a series of CPU write operations. Once initialized, the FDC is commanded to interact with the floppy disk drive (FDD). The FDD is the physical unit containing the motors, read/write head, and electronics needed to access the floppy disk. Figure II.9.3 shows the general connections between the floppy disk controller, the host CPU, and the floppy disk drive.

The FDC reacts to commands issued by the CPU and creates the appropriate signals for the FDD to function. These commands, for in-

Figure II.9.3 Basic FDC Connection
Courtesy of Intel Corp.

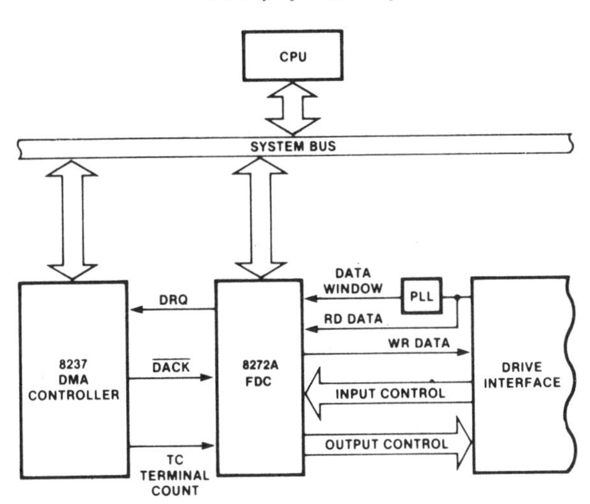

stance, may request that the FDC read data from a sector on a floppy disk.

The nature of magnetic recording necessitates some additional hardware between the FDC and the FDD. A Phased Locked Loop (PLL) is an integral part of a data separator circuit whose principal function is to translate magnetic pulse information into readable binary data. Therefore, all data read from the floppy disk passes through the PLL.

Since the floppy disk is often used as a storage device, it is highly probable that a quantity of data will be read/written at a time. Rather than access this data one byte at a time via the CPU, a DMA procedure can be used. The FDC has the standard signal lines used to respond to the DMA process initiated by an external DMA Controller. This option is specified at initialization time and can dramatically improve data throughput.

Read/Write Heads, Index Holes, and Other Floppy Disk Operations

During the execution of any FDC command, the mechanical elements of the floppy storage system are taken into consideration by the controller as they affect the sequence of data processing.

One of the first considerations is the Head Load and Unload Time. A floppy disk head reads or writes data when it is in direct contact with the spinning disk. This is referred to as Head Load and the length of time it takes the mechanical system to place the head in contact with the disk is called the Head Load Time. Contact is not always maintained to minimize both head and disk wear. Once contact is made, the head is not immediately ready to read or write data. The vibration caused by head and disk motion must diminish before a reliable data transfer can take place. A settling time for this motion is taken into account. This FDC can be programmed (using the Specify Command) for Head Load Time, which includes the time for head movement and settling (these parameters will vary, depending on the FDD used). The time programmed (2 to 254 msec in 2 msec steps) specifies the time between the Head Load (HDL) pin going high and the start of a read/write operation. The time to remove the head from contact, Head Unload time, can also be programmed in 16 msec increments from 16 msec to 240 msec. This time specifies when the head should be unloaded after the completion of a command. If another command is immediately issued before the head is unloaded, then the time required for the next command execution can be reduced since new Head Load and settling times are avoided. The Head Unload Time parameter thus helps minimize delay between repetitive disk drive operations.

In order to position the head over a particular track, a series of pulses must be delivered to the FDD motors controlling head movement. The direction of head movement is given to the FDD via the LCT/DIR (Low Current/Direction) line. The rate at which the step pulses occur is programmable (with the Specify Command) from 1 to 16 msec in 1 msec increments.

One of the floppy disk's accouterments, the index hole, aids the FDC in error detection. For instance, during a read or write the FDC will compare the recorded sector address to that specified in the currently executing command. The FDC reads all the sector addresses on the current track while searching for the correct one. If the address is not found after two revolutions of the disk (as measured by pulses generated from the index hole) a No Data condition is generated (bit 2 in Status Register 1). The processor becomes aware of this condition when reading the Status Registers during the Result Phase (discussed in The Programming Process section) of the command cycle.

Perhaps tracks, sectors, and addresses need clarification. The meanings that these definitions convey relate directly to the manner in which data is organized on the floppy disk surface. The word data in the last sentence is in itself a poor choice for describing the contents on the disk, since the computer user's definition of data describes only a small portion of the information ultimately placed on the magnetic face of the disk.

The physical aspects of sectors and tracks were discussed earlier, but, for the floppy disk to be a useful storage medium, the information must be reliably accessible in a software-based fashion. Since each track may have many sectors (programmable) and each disk many tracks, an addressing system must be in place to promote random access (Fig. II.9.4). Information carried at the beginning of each track (denoted by the presence of the index hole) includes an Index Address Mark and several gaps. The Index Address Mark uniquely identifies the track with a track address. This mark is read by the FDC to verify that the head is positioned above the proper track. The gaps are simply areas on the disk recorded with a fixed pattern. The purpose of the gap is to give some of the elements in the floppy disk system time to synchronize for proper operation.

Within each track are sectors that also must be identified. Therefore, each sector contains a Sector ID Field that not only identifies the sector but also contains information on the track, head, cylinder, and sector length. Note that this information is not data, but exists strictly

Figure II.9.4 Standard Floppy Diskette Track Format

Courtesy of Intel Corp.

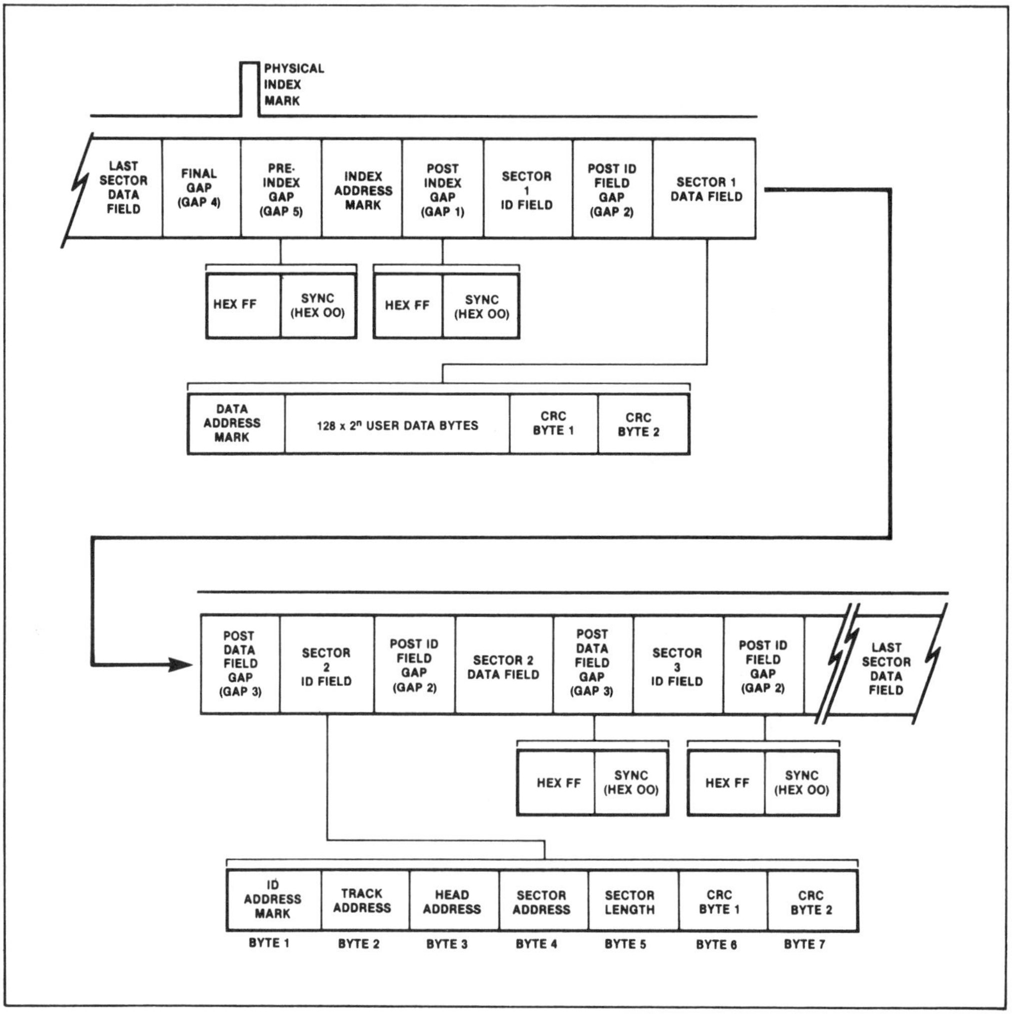

to allow the FDC to locate the data. Following the Sector ID and additional gap fields is the Data Field. This area on the disk is primarily for the data itself. At the beginning of the Data Field is another Address Mark that lets the FDC know that data is next (remember that the disk is spinning). The stored data is, typically 128, 256, 512, or 1024 bytes in length. The actual length is programmed during the Formatting process. After the data there is a two byte CRC (Cyclic Redundancy

Check) that ends the sector. The identification information for the following sector would be next on the track.

For the purpose of error checking, CRC bytes are generated for each sector's data and for each sector ID Field. The CRC is generated by the FDC when its corresponding information is written on the disk. The CRC is the result of a mathematical operation between a built-in polynomial and the stored data. (The CRC is actually the remainder of a division process between these two quantities.) The resulting CRC is always stored with the data used to generate it. In this system, data being read can be checked for errors when the FDC recomputes the CRC. If the recomputed CRC and the original CRC match, then the retrieved data was error free.

THE PROGRAMMING PROCESS

The FDC has a Command Set of fifteen commands that programs the controller for various operational modes and carries out specific FDD tasks. At the hardware level, the FDC is selected via a chip select line for Read or Write operation. A Read ($\overline{RD}$) or Write ($\overline{WR}$) line must be active in order for the appropriate event to take place. Much of the reading and writing will involve either the Main Status Register or the Data Register. The Main Status Register (Fig. II.9.5) is an 8-bit read-only register containing fundamental FDC information. The Data Register is also 8-bits wide and its information varies depending on the operational phase of the FDC. A signal line labeled A0 distinguishes between read and write operations to these two registers (A0 = 1 - Data Register, A0 = 0 - Status Register).

The manufacturers of the FDC break up command execution into three phases. The Command Phase occurs when the FDC is receiving information from the CPU. During this phase the CPU specifies one of four floppy disk drives for use. Further, it specifies recording modes and the overall operation to be performed (such as Read Data). This information requires the transfer of many bytes of data from the CPU to the FDC along the Data Bus. The sequence of transfer is precise and may not be altered. The FDC will expect a certain number of bytes in a specific order once a command is given (details discussed later). After all bytes are received, the FDC automatically switches to the Execution Phase of operation. In this phase, the FDC will carry out the operation it was commanded to do. The FDC will activate all necessary signal floppy disk drive lines during this time as well as interact with the CPU or DMA Controller if required. Once the command has been carried

Figure II.9.5 Main Status Register Bit Description
Courtesy of Intel Corp.

Main Status Register bit description.

BIT NUMBER	NAME	SYMBOL	DESCRIPTION
D_0	FDD 0 Busy	D_0B	FDD number 0 is in the Seek mode.
D_1	FDD 1 Busy	D_1B	FDD number 1 is in the Seek mode.
D_2	FDD 2 Busy	D_2B	FDD number 2 is in the Seek mode.
D_3	FDD 3 Busy	D_3B	FDD number 3 is in the Seek mode.
D_4	FDC Busy	CB	A read or write command is in process.
D_5	Non-DMA mode	NDM	The FDC is in the non-DMA mode. This bit is set only during the execution phase in non-DMA mode. Transition to "0" state indicates execution phase has ended.
D_6	Data Input/Output	DIO	Indicates direction of data transfer between FDC and Dta Register. If DIO = "1" then transfer is from Data Register to the Processor. If DIO = "0" then transfer is from the Processor to Data Register.
D_7	Request for Master	RQM	Indicates Data Register is ready to send or receive data to or from the Processor. Both bits DIO and RQM should be used to perform the handshaking functions of "ready" and "direction" to the processor.

out, the FDC enters the Result Phase. Status and other information is made available to the CPU at this time. The CPU must read this information or the Result Phase will not end. Again, the FDC is designed to expect a specific number of transfers to take place and will not move to another phase of operation until this has happened. After the Result Phase is complete, the FDC is ready for another command.

A Programming Example

The details of the programming process are even more exact than previously mentioned. A Command example will serve to elaborate on these fine points. Figure II.9.6 illustrates the Command Set of the FDC and Figure II.9.7 lists the mneumonics relevant to the commands. As is evident in Figure II.9.6, each command consists of a Command

Figure II.9.6 8272A Command Set

Courtesy of Intel Corp.

PHASE	R/W	DATA BUS								REMARKS
		D_7	D_6	D_5	D_4	D_3	D_2	D_1	D_0	
READ DATA										
Command	W	MT	MFM	SK	0	0	1	1	0	Command Codes
	W	0	0	0	0	0	HDS	DS1	DS0	
	W	C								Sector ID information prior to Command execution
	W	H								
	W	R								
	W	N								
	W	EOT								
	W	GPL								
	W	DTL								
Execution										Data transfer between the FDD and main-system
Result	R	ST 0								Status information after Command execution
	R	ST 1								
	R	ST 2								
	R	C								Sector ID information after command execution
	R	H								
	R	R								
	R	N								
READ DELETED DATA										
Command	W	MT	MFM	SK	0	1	1	0	0	Command Codes
	W	0	0	0	0	0	HDS	DS1	DS0	
	W	C								Sector ID information prior to Command execution
	W	H								
	W	R								
	W	N								
	W	EOT								
	W	GPL								
	W	DTL								
Execution										Data transfer between the FDD and main-system
Result	R	ST 0								Status information after Command execution
	R	ST 1								
	R	ST 2								
	R	C								Sector ID information after Command execution
	R	H								
	R	R								
	R	N								

PHASE	R/W	DATA BUS								REMARKS
		D_7	D_6	D_5	D_4	D_3	D_2	D_1	D_0	
WRITE DATA										
Command	W	MT	MFM	0	0	0	1	0	1	Command Codes
	W	0	0	0	0	0	HDS	DS1	DS0	
	W	C								Sector ID information prior to Command execution
	W	H								
	W	R								
	W	N								
	W	EOT								
	W	GPL								
	W	DTL								
Execution										Data transfer between the main-system and FDD
Result	R	ST 0								Status information after Command execution
	R	ST 1								
	R	ST 2								
	R	C								Sector ID information after Command execution
	R	H								
	R	R								
	R	N								
WRITE DELETED DATA										
Command	W	MT	MFM	0	0	1	0	0	1	Command Codes
	W	0	0	0	0	0	HDS	DS1	DS0	
	W	C								Sector ID information prior to Command execution
	W	H								
	W	R								
	W	N								
	W	EOT								
	W	GPL								
	W	DTL								
Execution										Data transfer between the FDD and main-system
Result	R	ST 0								Status information after Command execution
	R	ST 1								
	R	ST 2								
	R	C								Sector ID information after Command execution
	R	H								
	R	R								
	R	N								

Figure II.9.7 Command Mnemonics

Courtesy of Intel Corp.

READ A TRACK

PHASE	R/W	D_7	D_6	D_5	D_4	D_3	D_2	D_1	D_0	REMARKS
Command	W	0	MFM	SK	0	0	0	1	0	Command Codes
	W	0	0	0	0	0	HDS	DS1	DS0	
	W	C								Sector ID information prior to Command execution
	W	H								
	W	R								
	W	N								
	W	EOT								
	W	GPL								
	W	DTL								
Execution										Data transfer between the FDD and main-system. FDC reads all of cylinders contents from index hole to EOT
Result	R	ST 0								Status information after Command execution
	R	ST 1								
	R	ST 2								
	R	C								Sector ID information after Command execution
	R	H								
	R	R								
	R	N								

READ ID

PHASE	R/W	D_7	D_6	D_5	D_4	D_3	D_2	D_1	D_0	REMARKS
Command	W	0	MFM	0	0	1	0	1	0	Commands
	W	0	0	0	0	0	HDS	DS1	DS0	
Execution										The first correct ID information on the Cylinder is stored in Data Register
Result	R	ST 0								Status information after Command execution
	R	ST 1								
	R	ST 2								
	R	C								Sector ID information during Execution Phase
	R	H								
	R	R								
	R	N								

SCAN LOW OR EQUAL

PHASE	R/W	D_7	D_6	D_5	D_4	D_3	D_2	D_1	D_0	REMARKS
Command	W	MT	MFM	SK	1	1	0	0	1	Command Codes
	W	0	0	0	0	0	HDS	DS1	DS0	
	W	C								Sector ID information prior Command execution
	W	H								
	W	R								
	W	N								
	W	EOT								
	W	GPL								
	W	STP								
Execution										Data compared between the FDD and main-system
Result	R	ST 0								Status information after Command execution
	R	ST 1								
	R	ST 2								
	R	C								Sector ID information after Command execution
	R	H								
	R	R								
	R	N								

SCAN HIGH OR EQUAL

PHASE	R/W	D_7	D_6	D_5	D_4	D_3	D_2	D_1	D_0	REMARKS
Command	W	MT	MFM	SK	1	1	1	0	1	Command Codes
	W	0	0	0	0	0	HDS	DS1	DS0	
	W	C								Sector ID information prior Command execution
	W	H								
	W	R								
	W	N								
	W	EOT								
	W	GPL								
	W	STP								
Execution										Data compared between the FDD and main-system
Result	R	ST 0								Status information after Command execution
	R	ST 1								
	R	ST 2								
	R	C								Sector ID information after Command execution
	R	H								
	R	R								
	R	N								

FORMAT A TRACK

Phase	R/W	Data Bus	Remarks
Command	W	0 MFM 0 0 1 1 0 1	Command Codes
	W	0 0 0 0 0 HDS DS1 DS0	
	W	N	Bytes/Sector
	W	SC	Sectors/Cylinder
	W	GPL	Gap 3
	W	D	Filler Byte
Execution			FDC formats an entire cylinder
Result	R	ST 0	Status information after Command execution
	R	ST 1	
	R	ST 2	
	R	C	In this case, the ID information has no meaning
	R	H	
	R	R	
	R	N	

SCAN EQUAL

Phase	R/W	Data Bus	Remarks
Command	W	MT MFM SK 1 0 0 0 1	Command Codes
	W	0 0 0 0 0 HDS DS1 DS0	
	W	C	Sector ID information prior to Command execution
	W	H	
	W	R	
	W	N	
	W	EOT	
	W	GPL	
	W	STP	
Execution			Data compared between the FDD and main-system
Result	R	ST 0	Status information after Command execution
	R	ST 1	
	R	ST 2	
	R	C	Sector ID information after Command execution
	R	H	
	R	R	
	R	N	

RECALIBRATE

Phase	R/W	Data Bus	Remarks
Command	W	0 0 0 0 0 1 1 1	Command Codes
	W	0 0 0 0 0 0 DS1 DS0	
Execution			Head retracted to Track 0

SENSE INTERRUPT STATUS

Phase	R/W	Data Bus	Remarks
Command	W	0 0 0 0 1 0 0 0	Command Codes
Result	R	ST 0	Status information at the end of each seek operation about the FDC
	R	PCN	

SPECIFY

Phase	R/W	Data Bus	Remarks
Command	W	0 0 0 0 0 0 1 1	Command Codes
	W	SRT ← → HUT	
	W	HLT → ND	

SENSE DRIVE STATUS

Phase	R/W	Data Bus	Remarks
Command	W	0 0 0 0 0 1 0 0	Command Codes
	W	0 0 0 0 0 HDS DS1 DS0	
Result	R	ST 3	Status information about FDD

SEEK

Phase	R/W	Data Bus	Remarks
Command	W	0 0 0 0 1 1 1 1	Command Codes
	W	0 0 0 0 0 HDS DS1 DS0	
	W	NCN	
Execution			Head is positioned over proper Cylinder on Diskette

INVALID

Phase	R/W	Data Bus	Remarks
Command	W	Invalid Codes	Invalid Command Codes (NoOp — FDC goes into Standby State)
Result	R	ST 0	ST 0 = 80 (16)

(continued)

Figure II.9.7 Command Mnemonics (*cont.*)

SYMBOL	NAME	DESCRIPTION
A_0	Address Line 0	A_0 controls selection of Main Status Register ($A_0 = 0$) or Data Register ($A_0 = 1$).
C	Cylinder Number	C stands for the current selected Cylinder track number 0 through 76 of the medium.
D	Data	D stands for the data pattern which is going to be written into a Sector.
D_7–D_0	Data Bus	8-bit Data Bus where D_7 is the most significant bit, and D_0 is the least significant bit.
DS0, DS1	Drive Select	DS stands for a selected drive number 0 or 1.
DTL	Data Length	When N is defined as 00, DTL stands for the data length which users are going to read out or write into the Sector.
EOT	End of Track	EOT stands for the final Sector number of a Cylinder.
GPL	Gap Length	GPL stands for the length of Gap 3 (spacing between Sectors excluding VCO Sync Field).
H	Head Address	H stands for head number 0 or 1, as specified in ID field.
HDS	Head Select	HDS stands for a selected head number 0 or 1 (H = HDS in all command words).
HLT	Head Load Time	HLT stands for the head load time in the FDD (2 to 254 ms in 2 ms increments).
HUT	Head Unload Time	HUT stands for the head unload time after a read or write operation has occurred (16 to 240 ms in 16 ms increments).
MFM	FM or MFM Mode	If MF is low, FM mode is selected and if it is high, MFM mode is selected.
MT	Multi-Track	If MT is high, a multi-track operation is to be performed (a cylinder under both HD0 and HD1 will be read or written).
N	Number	N stands for the number of data bytes written in a Sector.

SYMBOL	NAME	DESCRIPTION
NCN	New Cylinder Number	NCN stands for a new Cylinder number, which is going to be reached as a result of the Seek operation. Desired position of Head.
ND	Non-DMA Mode	ND stands for operation in the Non-DMA Mode.
PCN	Present Cylinder Number	PCN stands for the Cylinder number at the completion of SENSE INTERRUPT STATUS Command. Position of Head at present time.
R	Record	R stands for the Sector number, which will be read or written.
R/W	Read/Write	R/W stands for either Read (R) or Write (W) signal.
SC	Sector	SC indicates the number of Sectors per Cylinder.
SK	Skip	SK stands for Skip Deleted Data Address Mark.
SRT	Step Rate Time	SRT stands for the Stepping Rate for the FDD (1 to 16 ms in 1 ms increments). The same Stepping Rate applies to all drives (F=1 ms, E=2 ms, etc.).
ST 0 ST 1 ST 2 ST 3	Status 0 Status 1 Status 2 Status 3	ST 0–3 stand for one of four registers which store the status information after a command has been executed. This information is available during the result phase after command execution. These registers should not be confused with the main status register (selected by $A_0 = 0$). ST 0–3 may be read only after a command has been executed and contain information relevant to that particular command.
STP		During a Scan operation, if STP = 1, the data in contiguous sectors is compared byte by byte with data sent from the processor (or DMA), and if STP = 2, then alternate sectors are read and compared.

Phase, an Execution Phase, and a Result Phase. The information opposite the Command and Result phases includes the data that must be read or written to properly carry out the instructions. For example, to read data from the FDD the Read Data Command must be issued. As a part of the Command Phase of the Read Data Command, nine bytes of data must be written to the FDC which lay the groundwork for the execution of the command. The five low order bits of the first byte (D4–D0 = 00110) designate that the command selected is Read Data. Bit D5 (SK), when high, informs the FDC to skip reading the sector with a Deleted Data Address Mark. The FDC will automatically read the next sector instead. Bit D6 (MFM) specifies the recording density used (MFM = 1 for MFM mode, MFM = 0 for FM mode). Bit D7 (MT), if high, informs the FDC that a double-sided disk is used. Figures II.9.8a and II.9.8b detail this byte and the remaining bytes of the Read Data Command.

The high order five bits of the second byte are all zeros. The remaining three bits, D2, D1, and D0 (HDS, DS1, DS0) are used to select the head (for double-sided disks) and FDD (one of four) used for the command.

The third byte is used to specify the cylinder (track) number (between 0 and 76) that contains the read data. The fourth byte will contain the number of the head and must match the head specified in bit D2 of the second byte.

Byte five indicates the sector number on the selected track for the read operation. The sixth byte tells the FDC the number of bytes to be found within the selected sector.

Byte seven, EOT, specifies the last sector number to be read. The eighth byte, GPL, denotes the Gap Length or spacing between sectors, and byte nine, DTL, is used for single sector reads of the data length.

All of this information is fed to the FDC as a sequence of write operations. What is not shown is that between every byte of write data, a read of the Main Status Register is required. Specifically, bits D6 and D7 (DIO and RQM) are checked. The register must be checked to assure that the processor and FDC are in sync with each other as far as the sequence of data transfers is concerned. The data associated with the command is written into the Data Register and, as mentioned, the precise sequence is critical. Bits D6 and D7 of the Main Status Register should be equal to 0 and 1 respectively during the Command Phase, indicating that the Data Register is ready to receive data from the processor.

Figure II.9.9 explains the exact procedure involved in properly carrying out the Command Phase of the Read Data Command. Since

Figure II.9.8 Read Data Command

	D_7	D_6	D_5	D_4	D_3	D_2	D_1	D_0
Byte 1 → Meaning →	MT 0 = Single Track 1 = Multi-track	MFM 1 = MFM 0 = FM	SK 1 = Skip 0 = No Skip	0 Read Data Op Code	0	1	1	0
Byte 2 → Meaning →	0 Always Zeros	0	0	0	0	HDS Head Select 0 or 1	DS1 DS0 Drive Select 00 = Drive 0 01 = Drive 1 10 = Drive 2 11 = Drive 3	
Byte 3 →	Cylinder/Track Number 0–76							
Byte 4 →	Head Address - Must Match Bit D2-Byte 2							
Byte 5 →	Sector Number							
Byte 6 →	Number of Data Byte per Sector							
Byte 7 →	End of Track - Final Sector Number on Track							
Byte 8 →	Gap Length of Post Data Field Gap							
Byte 9 →	Special Sector Size - Used to Alter Effective Sector Size							

(a) Command Phase of the Read Data Command

	D_7	D_6	D_5	D_4	D_3	D_2	D_1	D_0
Byte 1 →	Status Register 0 Contents							
Byte 2 →	Status Register 1 Contents							
Byte 3 →	Status Register 2 Contents							
Byte 4 →	Cylinder/Track Number at Command Termination							
Byte 5 →	Head Address at Command Termination							
Byte 6 →	Sector Number at Command Termination							
Byte 7 →	Data Bytes per Sector - Same as Specified in Command Phase							

(b) Result Phase of the Read Data Command

Figure II.9.9 Read Data Command Sequence (RDC)

READ DATA COMMAND SEQUENCE (RDC)

COMMAND PHASE
MAIN STATUS REG
BITS D6, D7 = 01
BEFORE A WRITE CAN OCCUR

RESULT PHASE
MAIN STATUS REG
BITS D6, D7 = 11
BEFORE A READ CAN OCCUR

PHASE	DATA BUS INFORMATION	A0	$\overline{RD}$	$\overline{WR}$
COMMAND PHASE ↓	READ MAIN STATUS REG	0	0	1
	WRITE BYTE 1 RDC	1	1	0
	READ MAIN STATUS REG	0	0	1
	WRITE BYTE 2 RDC	1	1	0
	READ MAIN STATUS REG	0	0	1
	WRITE BYTE 3 RDC	1	1	0
	READ MAIN STATUS REG	0	0	1
	WRITE BYTE 4 RDC	1	1	0
	READ MAIN STATUS REG	0	0	1
	WRITE BYTE 5 RDC	1	1	0
	READ MAIN STATUS REG	0	0	1
	WRITE BYTE 6 RDC	1	1	0
	READ MAIN STATUS REG	0	0	1
	WRITE BYTE 7 RDC	1	1	0
	READ MAIN STATUS REG	0	0	1
	WRITE BYTE 8 RDC	1	1	0
	READ MAIN STATUS REG	0	0	1
	WRITE BYTE 9 RDC	1	1	0
RESULT PHASE ↓	READ MAIN STATUS REG	0	0	1
	READ BYTE 1 RDC	1	0	1
	READ MAIN STATUS REG	0	0	1
	READ BYTE 2 RDC	1	0	1
	READ MAIN STATUS REG	0	0	1
	READ BYTE 3 RDC	1	0	1
	READ MAIN STATUS REG	0	0	1
	READ BYTE 4 RDC	1	0	1
	READ MAIN STATUS REG	0	0	1
	READ BYTE 5 RDC	1	0	1
	READ MAIN STATUS REG	0	0	1
	READ BYTE 6 RDC	1	0	1
	READ MAIN STATUS REG	0	0	1
	READ BYTE 7 RDC	1	0	1

there is really more than one Data Register, it is necessary to check thoroughly for Data Register availability. The Data Register is really several registers on a stack, but only one at a time is made available by the FDC. The checking just verifies that this complex sequence of events proceeds normally.

The FDC will automatically enter the Execution Phase once the last command byte is written (DTL for the Read Data Command). At this time the FDC provides the necessary signals which enable the FDD to carry out the desired operation. No CPU intervention is required during this stage of operation. When complete, the FDC then enters the Result Phase.

The Result Phase, like the Command Phase, involves the transfer of several bytes of data between the FDC and the CPU. In this instance, the CPU must read a certain number of bytes (depending on the command) to complete the whole command cycle. The bytes that are read include status information pertaining to the command execution that has just taken place. Once all the required bytes have been read the FDC returns to the beginning of the Command Phase to await the next CPU command. As before, between each byte read, the Main Status Register should be checked. For Result Phase operation, Main Status Register bits D6 and D7 should both be high, indicating that the Data Register is enabled to be read by the CPU. The actual bytes read for the Read Data Command are shown in Figure II.9.9.

The first three bytes contain information found in secondary Status Registers 0, 1, and 2. Figure II.9.10 details the meaning of the individual bits within these registers. Status Register 3 is also included in this illustration for completeness. Through these registers the success or reason for command failure can be determined. Many of the bits have meaning only for specific instructions. Following the status information, four additional bytes containing cylinder number, head address, sector number, and sector bytes must also be read. The information in these last four bytes is particularly meaningful in the event of an unsuccessful command execution.

The Remaining Commands

The Read Data Command is only one of fifteen supported by this FDC. The remaining fourteen are briefly explained here. As discussed, each command will proceed through the Command, Execution, and Result Phases. Refer to Figure II.9.6 for the Command and Result Phase bytes appropriate to a specific command.

Figure II.9.10 Status Registers

Courtesy of Intel Corp.

BIT			DESCRIPTION
NO.	NAME	SYMBOL	
			STATUS REGISTER 0
D_7	Interrupt Code	IC	$D_7 = 0$ and $D_6 = 0$ Normal Termination of Command, (NT). Command was completed and properly executed.
D_6			$D_7 = 0$ and $D_6 = 1$ Abnormal Termination of Command, (AT). Execution of Command was started, but was not successfully completed.
			$D_7 = 1$ and $D_6 = 0$ Invalid Command issue, (IC). Command which was issued was never started.
			$D_7 = 1$ and $D_6 = 1$ Abnormal Termination because during command execution the ready signal from FDD changed state.
D_5	Seek End	SE	When the FDC completes the SEEK Command, this flag is set to 1 (high).
D_4	Equipment Check	EC	If a fault Signal is received from the FDD, or if the Track 0 Signal fails to occur after 77 Step Pulses (Recalibrate Command) then this flag is set.
D_3	Not Ready	NR	When the FDD is in the not-ready state and a read or write command is issued, this flag is set. If a read or write command is issued to Side 1 of a single sided drive, then this flag is set.

BIT			DESCRIPTION
NO.	NAME	SYMBOL	
			STATUS REGISTER 1 (CONT.)
D_1	Not Writable	NW	During execution of WRITE DATA, WRITE DELETED DATA or Format A Cylinder Command, if the FDC detects a write protect signal from the FDD, then this flag is set.
D_0	Missing Address Mark	MA	If the FDC cannot detect the ID Address Mark after encountering the index hole twice, then this flag is set.
			If the FDC cannot detect the Data Address Mark or Deleted Data Address Mark, this flag is set. Also at the same time, the MD (Missing Address Mark in Data Field) of Status Register 2 is set.
			STATUS REGISTER 2
D_7			Not used. This bit is always 0 (low).
D_6	Control Mark	CM	During executing the READ DATA or SCAN Command, if the FDC encounters a Sector which contains a Deleted Data Address Mark, this flag is set.
D_5	Data Error in Data Field	DD	If the FDC detects a CRC error in the data field then this flag is set.
D_4	Wrong Cylinder	WC	This bit is related with the ND bit, and when the contents of C on the medium is different from that stored in the IDR, this flag is set.
D_3	Scan Equal Hit	SH	During execution, the SCAN Command, if the condition of "equal" is satisfied, this flag is set.

D_2	Head Address	HD	This flag is used to indicate the state of the head at Interrupt.
D_1	Unit Select 1	US 1	These flags are used to indicate a Drive Unit Number at Interrupt
D_0	Unit Select 0	US 0	
	STATUS REGISTER 1		
D_7	End of Cylinder	EN	When the FDC tries to access a Sector beyond the final Sector of a Cylinder, this flag is set.
D_6			Not used. This bit is always 0 (low).
D_5	Data Error	DE	When the FDC detects a CRC error in either the ID field or the data field, this flag is set.
D_4	Over Run	OR	If the FDC is not serviced by the main-systems during data transfers, within a certain time interval, this flag is set.
D_3			Not used. This bit always 0 (low).
D_2	No Data	ND	During execution of READ DATA, WRITE DELETED DATA or SCAN Command, if the FDC cannot find the Sector specified in the IDR Register, this flag is set.
			During executing the READ ID Command, if the FDC cannot read the ID field without an error, then this flag is set.
			During the execution of the READ A Cylinder Command, if the starting sector cannot be found, then this flag is set.

D_2	Scan Not Satisfied	SN	During executing the SCAN Command, if the FDC cannot find a Sector on the cylinder which meets the condition, then this flag is set.
D_1	Bad Cylinder	BC	This bit is related with the ND bit, and when the content of C on the medium is different from that stored in the IDR and the content of C is FF, then this flag is set.
D_0	Missing Address Mark in Data Field	MD	When data is read from the medium, if the FDC cannot find a Data Address Mark or Deleted Data Address Mark, then this flag is set.
	STATUS REGISTER 3		
D_7	Fault	FT	This bit is used to indicate the status of the Fault signal from the FDD.
D_6	Write Protected	WP	This bit is used to indicate the status of the Write Protected signal from the FDD.
D_5	Ready	RDY	This bit is used to indicate the status of the Ready signal from the FDD.
D_4	Track 0	T0	This bit is used to indicate the status of the Track 0 signal from the FDD.
D_3	Two Side	TS	This bit is used to indicate the status of the Two Side signal from the FDD.
D_2	Head Address	HD	This bit is used to indicate the status of Side Select signal to the FDD.
D_1	Unit Select 1	US 1	This bit is used to indicate the status of the Unit Select 1 signal to the FDD.
D_0	Unit Select 0	US 0	This bit is used to indicate the status of the Unit Select 0 signal to the FDD.

- *Write Data.* This command is basically the opposite of the Read Data Command. The FDC is instructed to write data to a specified location on the disk and will continue with the operation until signaled to stop. The Terminal Count signal gives this indication. Errors, such as overrun and incorrect CRC, are checked.
- *Read Deleted Data.* Data sectors that are unusable due to hard errors on the disk surface are marked with a special address mark called a Deleted Data Address Mark. The Read Deleted Data Command, which is very similar to Read Data, can be used to read these bad sectors. Two options are possible. When the skip flag (SK) in the first Command byte is not set (meaning do not skip this sector), all the data in the sector with the Deleted Data Address Mark is read, the Control Mark (CM) bit in Status Register 2 is set, and the command is terminated. When SK is set, the bad sector is skipped and the next sector is read.
- *Write Deleted Data.* This command marks bad sectors on the floppy disk by writing a special Deleted Data Address Mark in the beginning of the Data Field.
- *Read a Track.* All data sectors on a track are read as continuous data. The FDC begins reading after detecting the Index Hole. Errors occur if the Address ID specified in the Command Phase is not found on the track. The No Data (ND) flag in Status Register 1 is set under these circumstances. After two passes over the track without finding the Address ID, the Missing Address (MA) flag is set. CRC checks are also performed.
- *Read ID.* Read ID returns the position of the read/write head. The FDC reads the first ID Field it encounters. MA and ND flags in Status Register 1 are set if no valid Address ID is found within two revolutions of the disk.
- *Format a Track.* An entire track can be formatted with this command according to the IBM System 34 (DD) or IBM 3740 (SD) recording formats. The Command Phase data dictates the format by specifying the number of bytes per sector, gap lengths, number of sectors per cylinder, and the data byte pattern used. The CPU will be requested to send cylinder, head, sector, and bytes/sector information after each sector on the track is formatted. Thus, the CPU provides the Address ID information for each sector. This method allows for nonsequential sector numbering, called interleaving, which can optimize access times.

- *Scan Equal.* This command, as well as the next two scan commands, provides the programmer with the means to compare data from the floppy disk with data from the computer system. The data is compared a byte at a time until a comparison occurs, the end of the track is reached, or a terminating signal is received. The Scan Not Satisfied (SN) and Scan Hit (SH) bits of Status Register 2 (bits 2 and 3 respectively) indicate the results of the scan commands. For the Scan Equal Command, bits 2,3 = 0,1 means a comparison occurred. Bits 2,3 = 1,0, implies that no compare occurred.
- *Scan Low or Equal.* This scan command can compare floppy disk data and processor data for conditions of disk data less than or equal to processor data. Results are indicated in Status Register 2. Bits 2,3 = 0,1 for a comparison; bits 2,3 = 0,0 when floppy disk data is less than processor data; bits 2,3 = 1,0 when the floppy disk data is greater than the processor data.
- *Scan High or Equal.* The third scan command can compare floppy disk data and processor data for conditions of disk data greater than or equal to processor data. Results are indicated in Status Register 2. Bits 2,3 = 0,1 for a comparison; bits 2,3 = 0,0 when floppy disk data is greater than processor data; bits 2,3 = 1,0 when the floppy disk data is less than the processor data.
- *Recalibrate.* To recalibrate means to place the read/write head over Track 0 (outermost track on the floppy disk). This command causes this response to take place and sets up the FDC to move the read/write head toward the center of the disk on the next read or write action (Low Current/Direction = 1). Pin 33 (Fault/Track 0) goes high when Track 0 is encountered. Furthermore, the Present Cylinder Number (PCN) Register is cleared and the Seek End (SE) flag in Status Register 0 is set indicating that the head is over Track 0. If, for some reason, the Track 0 line remains low after 77 step pulses have been issued, an error is signaled. The command is ended with both SE and EC (Equipment Check) set high. Note, there is no Result Phase for this command.
- *Seek.* The Seek Command is used to position the read/write head over the desired track for disk read and write operations. Two internal registers are involved. The Present Cylinder Number (PCN) Register contains the current head position. The New Cylinder Number (NCN) Register is programmed during the

Command Phase with the track address desired. During Execution Phase, the FDC steps the read/write head toward the requested cylinder/track at a rate preprogrammed. After each step, the PCN and NCN registers are compared. Pin 38 (Low Current/Direction) is set to one when PCN < NCN or set to zero when PCN > NCN. This determines the direction of head travel. When PCN equals NCN the proper track has been reached. Note, there is no Result Phase for this command. Status Register bit SE is set high to indicate comparison and the command ends. Since all read and write operations require the proper track to be located preceding the read or write, the following sequence is recommended: Issue the Seek Command, specifying the desired track. Issue a Sense Interrupt Status Command (described below). Issue a Read ID Command. If these steps verify that the proper track has been reached, then issue the read or write command.

- *Sense Interrupt Status.* Many interrupts can be generated using this FDC. The CPU handles interrupts with special sets of software that determine and act upon the reason for the interrupt. Some interrupts are not easily discerned by the CPU (such as those caused by the Seek or Recalibrate Commands). The Sense Interrupt Status Command returns the reason for a previous interrupt during the Result Phase in Status Register 0, bits 5, 6, and 7.
- *Sense Drive Status.* Through this command, Status Register 3 will return the status of the selected drive.
- *Specify.* This command is typically one of the first to be executed. Specify is used to program values for Head Unload Time, Step Rate Time, Head Load Time, and the choice of DMA or non-DMA operation. There is no Result Phase from the Specify Command.

Finally, if an invalid command is sent to the FDC the FDC will go into a stand-by state of operation. Status Register 0 will contain a 80H, indicating that an invalid command was received. No interrupt is generated when the invalid command is terminated.

DATA TRANSFER MODES

The two modes of data transfer through this FDC are referred to as DMA and non-DMA. The Specify Command is used to select the mode.

In DMA mode, an associated DMA Controller chip is used to transfer data to and from the FDC using DMA techniques. An active Terminal Count (TC) signal from the DMA Controller terminates the execution of FDC commands.

Non-DMA mode uses interrupts to control the transfer of data (note, interrupts are still used in DMA mode to signal the ends of commands and errors). In this mode the CPU controls the Terminal Count signal, activating it when the proper amount of data has been transferred. The choice of mode may be selected with regard to the service time required by the FDC. When using FM format, the FDD must be serviced by the processor or DMA Controller every 27 μsec. Servicing is required every 13 μsec using MFM format.

It should be evident that this Floppy Disk Controller is a complex device which can greatly simplify the hardware design aspects of a floppy disk system. However, a great deal of intelligent software is also required to issue FDC commands while interacting with the computer system as a whole. For example, an extensive part of the software will be used to gracefully recover from what seems like an indeterminable number of potential error situations. Clearly, the hardware and software designer have their work cut out for them.

II.10

CMOS Real-Time Clock with RAM

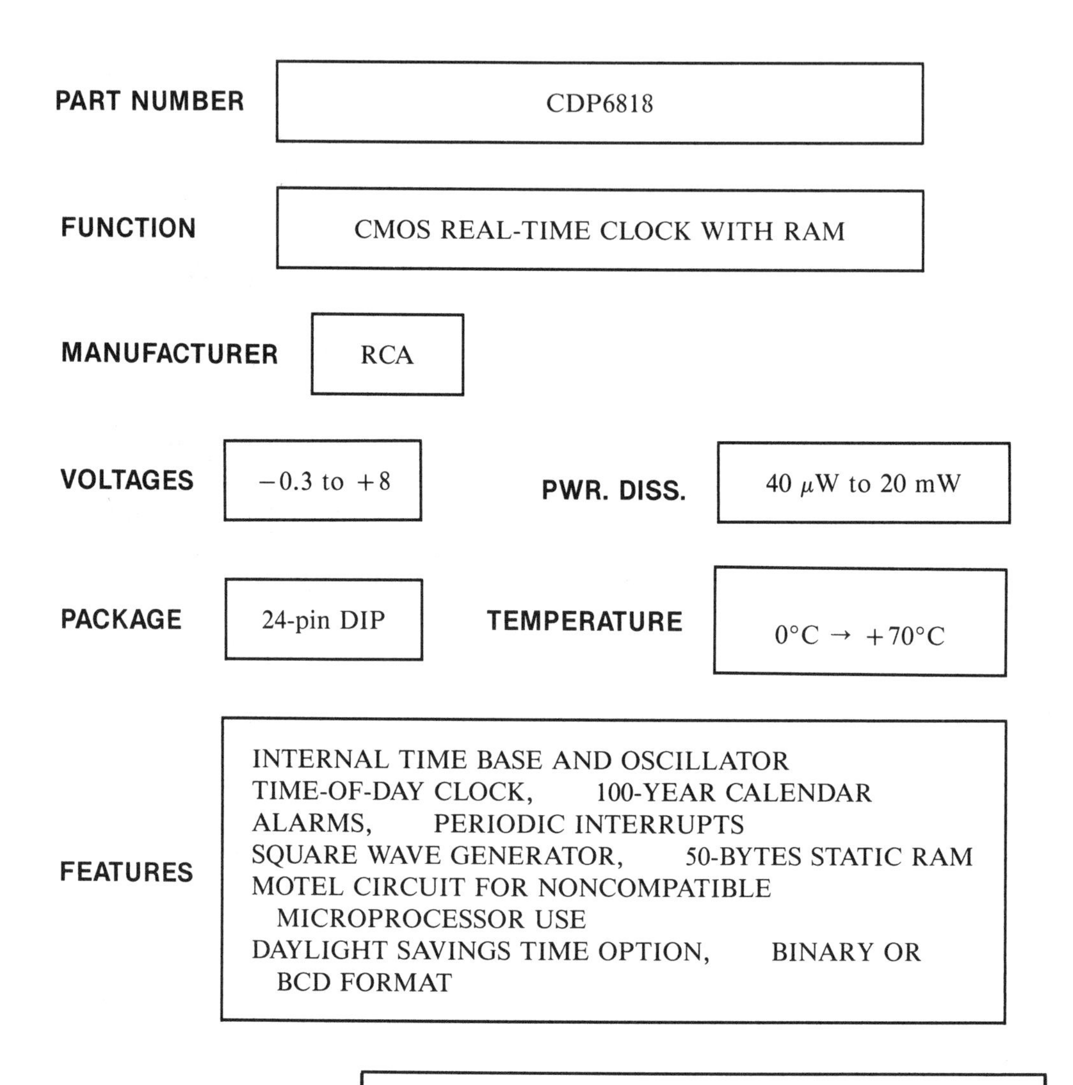

PART NUMBER	CDP6818
FUNCTION	CMOS REAL-TIME CLOCK WITH RAM
MANUFACTURER	RCA
VOLTAGES	−0.3 to +8
PWR. DISS.	40 μW to 20 mW
PACKAGE	24-pin DIP
TEMPERATURE	0°C → +70°C

FEATURES

INTERNAL TIME BASE AND OSCILLATOR
TIME-OF-DAY CLOCK, 100-YEAR CALENDAR
ALARMS, PERIODIC INTERRUPTS
SQUARE WAVE GENERATOR, 50-BYTES STATIC RAM
MOTEL CIRCUIT FOR NONCOMPATIBLE MICROPROCESSOR USE
DAYLIGHT SAVINGS TIME OPTION, BINARY OR BCD FORMAT

COMPATIBLE MICROPROCESSORS

6800 FAMILY-EXTENDED TO OTHER PROCESSOR FAMILIES VIA THE MOTEL CIRCUIT

FUNCTIONAL DESCRIPTION

REAL-TIME CLOCK AND CALENDAR FUNCTIONS ARE POSSIBLE WITH THIS CHIP. TIME IS KEPT IN RAM LOCATIONS AND PERIODICALLY UPDATED BY THE CHIP WITHOUT PROCESSOR INTERVENTION. ALARMS AND TIME OUTS CAN BE SENSED BY INTERRUPTS OR POLLING. ON-CHIP RAM AND POWER SENSING, COMBINED WITH BATTERY POWERED OPERATION, PROVIDE BACKUP MEMORY CAPABILITY.

Figure II.10.1 CDP6818 Pinout
Courtesy of GE Solid State

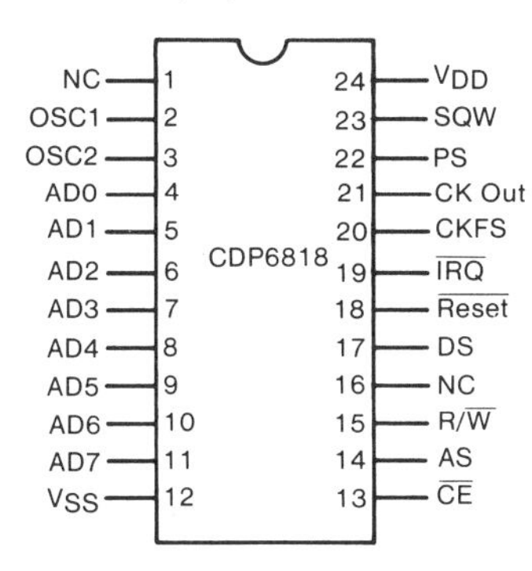

Figure II.10.2 CDP6818 Block Diagram
Courtesy of GE Solid State

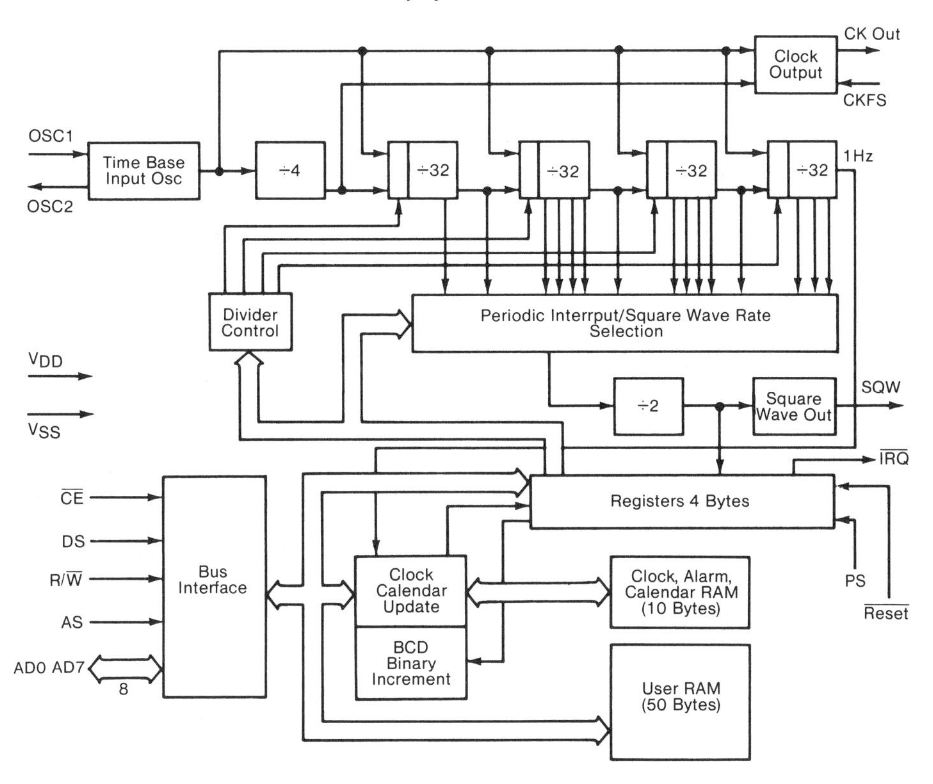

PIN NAME	PIN SYMBOL	PIN NUMBER	FUNCTION
OSCILLATOR 1	OSC 1	2	This input controls the chip timing. The timing signal source may be in the form of an externally generated square wave applied to this pin or produced from a crystal. The square wave frequencies can be either 4.194304 MHz, 1.048576 MHz, or 32.768 kHz.
OSCILLATOR 2	OSC 2	3	If a crystal is used for timing control, one lead of the crystal will be connected here. An AT-cut crystal of frequencies 4.194304 MHz or 1.048576 MHz is recommended. The on chip oscillator also requires that the crystal be specified for parallel resonant mode of operation.
ADDRESS/DATA BUS 0	AD0	4	These lines comprise a multiplexed address and data bus. On this bus, initialization information and the exchange of data between the CDP6818 and the host CPU take place.
ADDRESS/DATA BUS 1	AD1	5	
ADDRESS/DATA BUS 2	AD2	6	
ADDRESS/DATA BUS 3	AD3	7	
ADDRESS/DATA BUS 4	AD4	8	
ADDRESS/DATA BUS 5	AD5	9	
ADDRESS/DATA BUS 6	AD6	10	
ADDRESS/DATA BUS 7	AD7	11	

PIN NAME	PIN SYMBOL	PIN NUMBER	FUNCTION
CHIP ENABLE	$\overline{CE}$	13	The Real-Time Clock (RTC) can be accessed by the host CPU only when chip enable is low. When $\overline{CE}$ is high, the multiplexed outputs are placed into a high impedance state. Furthermore, inputs to the CDP6818 are isolated from the CPU under these conditions to conserve battery power.
ADDRESS STROBE	AS	14	Address strobe is used to latch address information from the address/data bus. The CDP6818 contains internal latches for this purpose. This control line is obtained from the CPU.
READ/WRITE	R/$\overline{W}$	15	This line, from a compatible CPU (6800 type), indicates that a read or write is taking place. For noncompatible processors, an internal circuit (MOTEL) will treat this input as a low active write pulse.
DATA STROBE	DS	17	With a compatible processor, DS indicates when the RTC should place data on the address/data bus or when to latch data present on the bus. The MOTEL circuit translates this signal into a negative active read ($\overline{RD}$) line.

(*continued*)

PIN NAME	PIN SYMBOL	PIN NUMBER	FUNCTION
RESET	$\overline{RESET}$	18	A low level on the reset input for 5 μsec minimum will clear a number of RTC individual register bits to zero. See text for details.
INTERRUPT REQUEST	$\overline{IRQ}$	19	This output line, when low, indicates a pending interrupt. The line is made inactive by either a reset or by reading Register C.
CLOCK OUT FREQUENCY SELECT	CKFS	20	This input controls the Clock Out pin. When CKFS is tied to Vdd, Clock Out will equal the frequency at the Osc 1 pin. When CKFS is tied to Vss, Clock Out will equal the Osc 1 frequency divided by four.
CLOCK OUT	CK OUT	21	This output is related to the OSC 1 input frequency. The exact frequency depends on the frequency present at the Osc 1 input and the voltage level at the CKFS input. (See above). This output can drive the microprocessor clock input, eliminating the need for an extra crystal.
POWER SENSE	PS	22	This line is normally combined with an external RC network to validate RAM contents during the power-on process.

PIN NAME	PIN SYMBOL	PIN NUMBER	FUNCTION
SQUARE WAVE	SQW	23	A square wave can be programmed to appear at this output. Fifteen possible frequency selections are available by programming.
Vdd	Vdd	24	Supply voltage—typically +5 V.
Vss	Vss	12	Typically 0 V.
NO CONNECTION	NC	1	These pins are not used.
NO CONNECTION	NC	16	

Real-Time Clocks (RTC) keep track of items such as current time and date in a computer system. RTCs are separate pieces of hardware dedicated to this task. The resultant time information is often used to time-stamp data files or output information. In many of the popular microcomputer systems available, Real-Time Clock cards are used for this function. The user simply plugs the card into an available slot in his system to make the logical connection. The CDP6818 (Figs. II.10.1 and II.10.2) is a single chip design that allows an RTC to be easily added to a microprocessor system. For applications that demand Time-of-Day clock or calendar functions, the CDP6818 offers a useful combination of time, date, and alarm capabilities with the added bonus of RAM storage. The CMOS designed chip can be run with battery power, making the RAM ideal for loss of power backup storage. In fact, the chip is designed to sense loss of power so that critical timekeeping functions can remain operational and unaffected even with the electrical disruptions that occur as power is switched over to batteries.

THE MICROPROCESSOR CONNECTION

The RTC is treated as a block of memory by the controlling microprocessor (MPU). This block of memory is 64 bytes in size divided into several functional components. Fifty bytes are general purpose RAM. Ten bytes are used to keep track of time, date, and alarms while the remaining four bytes are treated as control registers. Read and write access to this memory block takes place through the microprocessor bus. The chip has been designed to work with a multiplexed address and data bus. The CDP6818 is compatible with 6800 family processors, but the inclusion of an interfacing circuit on the chip known as MOTEL allows the RTC to be easily connected with other multiplexed bus MPUs (such as the 8085). Figure II.10.3 shows how several control lines to the CDP6818 are used to interface with non-6800 microprocessors.

Data and addresses are transferred between the RTC and the MPU over the bus using normal memory read and write instructions. Address information provided during instruction execution is internally latched within the RTC. The Address Strobe (AS) line is connected to the Address Strobe or Address Latch Enable (ALE) pin on the MPU to coordinate the acceptance of address information. In the typical read/write cycle, data follows the address information on the bus. The Read/Write ($R/\overline{W}$) line indicates the type of operation that will occur, while the Data Strobe (DS) line indicates the actual data transfer timing. If an MPU other than a 6800 type is used, then Data Strobe and $R/\overline{W}$ have different meaning through the MOTEL circuitry. $R/\overline{W}$ becomes the write input and Data Strobe becomes the Read input. Refer again to Figure II.10.3. Since this interface is transparent to the user of the RTC, the connection to the desired MPU is easily accomplished.

With any peripheral that shares the data bus, a selection process must take place to allow the device complete access to the data bus resource. Chip Enable ($\overline{CE}$), when asserted low, enables the transfer of information between MPU and RTC to take place. When $\overline{CE}$ is high, internal timekeeping functions continue, although the RTC output bus is placed in a high impedance state.

Figure II.10.3 Functional Diagram of MOTEL Circuit

Courtesy of GE Solid State

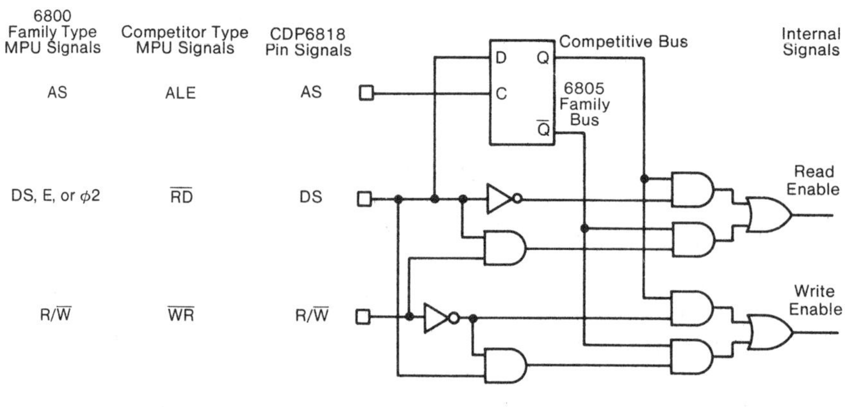

The Timing Process

The memory map (Fig. II.10.4) indicates the ten bytes used to contain time and date information. A program that needs this information reads the appropriate memory location to obtain it. The time information stored in these locations is a result of previous write op-

Figure II.10.4(a) RTC Memory Map

Courtesy of GE Solid State

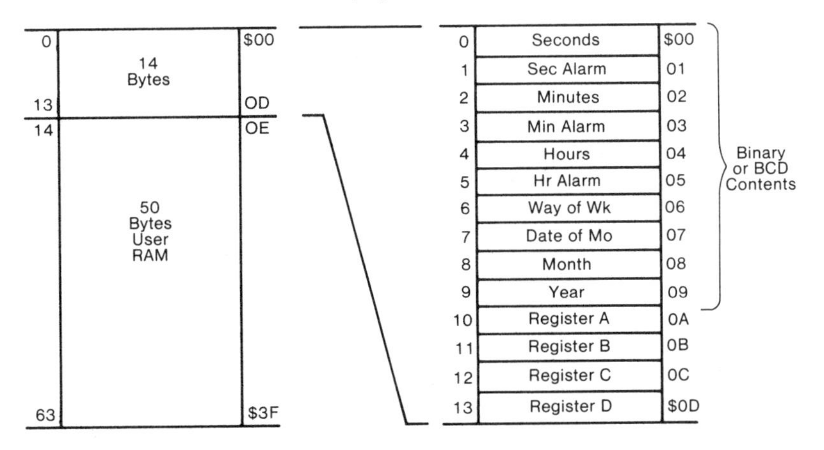

Figure II.10.4(b) Time and Date Representation in the RTC

Courtesy of GE Solid State

Address Location	Function	Decimal Range	Range		Example*	
			Binary Data Mode	BCD Data Mode	Binary Data Mode	BCD Data Mode
0	Seconds	0-59	$00-$3B	$00-$59	15	21
1	Seconds Alarm	0-59	$00-$3B	$00-$59	15	21
2	Minutes	0-59	$00-$3B	$00-$59	3A	58
3	Minutes Alarm	0-59	$00-$3B	$00-$59	3A	58
4	Hours (12 Hour Mode)	1-12	$01-$0C (AM) and $81-$8C (PM)	$01-$12 (AM) and $81-$92 (PM)	05	05
	Hours (24 Hour Mode)	0-23	$00-$17	$00-$23	05	05
5	Hours Alarm (12 Hour Mode)	1-12	$01-$0C (AM) and $81-$8C (PM)	$01-$12 (AM) and $81-$92 (PM)	05	05
	Hours Alarm (24 Hour Mode)	0-23	$00-$17	$00-23	05	05
6	Day of the Week Sunday = 1	1-7	$01-$ 07	$01-$07	05	05
7	Day of the Month	1-31	$01-$1F	$01-$31	0F	15
8	Month	1-12	$01-$0C	$01-$12	02	02
9	Year	0-99	$00-$63	$00-$99	4F	79

*Example: 5:58:21 Thursday February 15 1979 (Time is A.M.)

erations. The contents of these ten bytes can be represented in either binary or BCD form as specified by the data mode (DM) bit in register B. The chart in Figure II.10.4b provides the numerical ranges used to represent time and date information in the RTC. From this chart it can be noted that time may be represented in either 12- or 24-hour modes (i.e., 1 P.M. is 13 hours in 24-hour mode). The 12/24 bit in Register B sets this mode.

As mentioned previously, an initial write must be made to the seconds, minutes, hours, days of week, dates of month, months, and years bytes to initialize them. Once every second thereafter, the RTC updates these bytes, if necessary, to keep track of the current time and date. To prevent an update from accidentally occurring during the initialization procedure, another bit in Register B (SET) can be used to suppress the normal updating process.

The alarm bytes are also initialized. For instance, if an indication is necessary when 3 A.M. is reached, then the hours alarm byte would be set to 3 A.M. The RTC compares the hours byte with the hours alarm byte and, when a match occurs, sends an interrupt to the MPU on the Interrupt Request ($\overline{IRQ}$) line. A similar procedure is followed for the seconds and minutes bytes. In addition, a don't-care code (11XXXXXX) can be placed in any of the alarm bytes. In this case, an interrupt will be raised every second, minute, or hour. The alarms may all be disabled through the alarm interrupt enable bit (AIE) in Register B.

Once initialized and allowed to run, the RTC will accurately keep track of the time and date. The programs running on the MPU can reference the RTC information by reading the desired memory locations. Since the RTC must update the time and date information on a periodic basis, it is possible that an update may be in progress when a program attempts to read time information. During an update, the time and date information is not made available to the system bus. Therefore, a read to these locations would result in erroneous data. Several methods can be used to avoid reading incorrect time and date information. Two methods involve interrupts (see the Registers section) and a third involves checking a status bit. Register A contains a bit called the update-in-progress bit (UIP). When this bit goes high, 244 μsec remain until the next update occurs. Checking this bit will indicate if an update is in progress and, if the bit is low, indicates that at least 244 μsec remain before the next update takes place. Data can safely be read during this time span. Incidentally, the length of time necessary for the RTC to perform an update depends upon the clock frequency driving the chip. A 4.194304 MHz or 1.048576 MHz time base will produce an update time of 248 μsec, while the 32.768 kHz time base results in an update time of 1984 μsec. Time and calendar information are not assured during these update times.

CDP6818 Registers

Four of the RAM locations in the CDP6818 are referred to as Registers A, B, C, and D. These Registers control the various options available within the Real-Time Clock chip. The Registers can be read and written like any other location in the chip, even when a timing update cycle is in progress. The individual bits within each register control many RTC functions and are discussed below.

- *Register A*. The majority of the bits in this register affect the timing elements of the chip (Fig. II.10.5). The four low order bits are labeled RS0, RS1, RS2, and RS3. These bits will disable

Figure II.10.5 CDP6818 Registers
Courtesy of GE Solid State

Register A - OA

B7	B6	B5	B4	B3	B2	B1	B0	Bit
UIP	DV2	DV1	DV0	RS3	RS2	RS1	RS0	Function

Register B - OB

B7	B6	B5	B4	B3	B2	B1	B0	Bit
SET	PIE	AIE	UIE	SQWE	DM	24/12	DSE	Function

Register C - OC

B7	B6	B5	B4	B3	B2	B1	B0	Bit
IRQF	PF	AF	UF	0	0	0	0	Function

Register - OD

B7	B6	B5	B4	B3	B2	B1	B0	Bit
VRT	0	0	0	0	0	0	0	Function

CDP 6818 Registers

or set the frequency rate for periodic interrupts and the square wave output pin. The possibilities are shown in Figure II.10.6. These output frequencies are derived from a 22-stage divider that runs off the time base oscillator.

Bits 4, 5, and 6 (DV0, DV1, DV2) select the time base frequency being used. The bits are also used to reset the 22-stage divider as well as to assure an accurate start to the Time-of-Day Clock after initialization. Assume that Register A bits DV2, DV1, and DV0 are set to 110. This combination will reset the divider chain. Further assume that the time base oscillator is set to run at 1.048576 MHz. Through another write instruction (such as an STA), DV2, DV1, and DV0 can be changed to 001. According to Figure II.10.7, this combination will place the chip in operation. The actual timing will begin exactly one-half second after the chip is made operational by the command mentioned.

The most significant bit is labeled UIP for update-in-progress. This is a status flag (read only) that is set to a high level by the chip when an update is currently in progress. During

Figure II.10.6 Periodic Interrupt Rate and Square Wave Output Frequency
Courtesy of GE Solid State

PERIODIC INTERRUPT RATE AND SQUARE WAVE OUTPUT FREQUENCY

Rate Select Control Register A				4.194304 or 1.048576 MHz Time Base		32.768 kHz Time Base	
RS3	RS2	RS1	RS0	Periodic Interrupt Rate t_{PI}	SQW Output Frequency	Periodic Interrupt Rate t_{PI}	SQW Output Frequency
0	0	0	0	None	None	None	None
0	0	0	1	30 517 μs	32 768 kHz	3 90625 ms	256 Hz
0	0	1	0	61 035 μs	16 384 kHz	7 8125 ms	128 Hz
0	0	1	1	122 070 μs	8 192 kHz	122 070 μs	8 192 kHz
0	1	0	0	244 141 μs	4 096 kHz	244 141 μs	4 096 kHz
0	1	0	1	488 281 μs	2 048 kHz	488 281 μs	2 048 kHz
0	1	1	0	976 562 μs	1 024 kHz	976 562 μs	1 024 kHz
0	1	1	.1	1 953125 ms	512 Hz	1 953125 ms	512 Hz
1	0	0	0	3 90625 ms	256 Hz	3 90625 ms	256 Hz
1	0	0	1	7 8125 ms	128 Hz	7 8125 ms	128 Hz
1	0	1	0	15 625 ms	64 Hz	15 625 ms	64 Hz
1	0	1	1	31 25 ms	32 Hz	31 25 ms	32 Hz
1	1	0	0	62 5 ms	16 Hz	62 5 ms	16 Hz
1	1	0	1	125 ms	8 Hz	125 ms	8 Hz
1	1	1	0	250 ms	4 Hz	250 ms	4 Hz
1	1	1	1	500 ms	2 Hz	500 ms	2 Hz

Figure II.10.7 Divider Configurations
Courtesy of GE Solid State

DIVIDER CONFIGURATIONS

Time-Base Frequency	Divider Bits Register A			Operation Mode	Divider Reset	Bypass First N-Divider Bits
	DV2	DV1	DV0			
4.194304 MHz	0	0	0	Yes		N = 0
1.048576 MHz	0	0	1	Yes		N = 2
32.768 kHz	0	1	0	Yes		N = 7
Any	1	1	0	No	Yes	
Any	1	1	1	No	Yes	

Note: Other combinations of divider bits are used for test purposes only.

this update time, time and calendar information are not available to the MPU.

None of the bits in Register A are affected by a RESET.

- *Register B.* Register B controls many of the RTC interrupts and modes of operation. Each bit in this register is set aside for a specific function and is typically initialized prior to the start of timekeeping operations. The least significant bit is the daylight saving time enable (DSE) bit and is used to automatically compensate for the changes in time that occur in the spring and fall. When this bit is set to "1," the update for the daylight saving time function is enabled. The second bit (B1) is labeled 24/12. When set to "1," time is kept in 24-hour mode. The time and calendar data can be represented in binary or BCD formats. The data mode (DM) bit (B2) selects BCD if it is low. Bit B3 controls the square-wave output pin. A square-wave is produced at a frequency determined by RS0 through RS3 in Register A if bit B3 (SQWE) is set to one. A zero in this bit position will bring the square-wave output pin to a low level. Unlike the bits previously discussed, this bit is affected by a $\overline{\text{RESET}}$. A $\overline{\text{RESET}}$ clears the SQWE bit.

  Three of the bits in register B are devoted to enabling the various types of interrupts supported by the RTC chip. Each enable bit has a corresponding flag bit in Register C that will, if enabled (with a high level), assert the interrupt request line ($\overline{\text{IRQ}}$). Bit B4 is the update-ended, interrupt enable (UIE) bit and is used to enable an interrupt whenever the timekeeping update cycle has ended. Bit B5 (AIE) enables the alarm interrupt as described earlier, while PIE (B6) enables periodic interrupts. A periodic interrupt can be generated according to the rate specified by bits RS0 through RS3 in Register A (Fig. II.10.6). The three enable bits are all cleared to zero by a $\overline{\text{RESET}}$. The UIE bit is also cleared by setting the SET bit (B7) high. SET is brought to the high state to suppress the time and calendar update during the initialization process. SET is not affected by a $\overline{\text{RESET}}$.

- *Register C.* Only the four most significant bits of this register are used. The four low order bits are permanently held at the low level. Bit B7 (interrupt request flag IRFQ) reflects the state of the interrupt request pin ($\overline{\text{IRQ}}$). When this bit is set high by the RTC chip, an interrupt is generated ($\overline{\text{IRQ}}$ = 0). The remaining three bits in the register are flags for specific interrupts.

  The periodic interrupt flag (PF) attains a high level at a predetermined rate specified by bits RS0 through RS3 in Register A. If the periodic interrupt enable bit (PIE) in Register A is set high, an interrupt will be generated ($\overline{\text{IRQ}}$ = 0 and IRQF = 1) each time the selected periodic interrupt rate time has elapsed. This type of interrupt is useful for initiating processes that must occur at a regular rate.

  The alarm interrupt flag (AF) is raised high by the RTC chip whenever an alarm byte matches its corresponding time byte or don't-care byte (11XXXXXX). The use of don't-care bytes is another mechanism whereby a periodic interrupt can be generated. In any case, the interrupt will be generated only if the AIE (alarm interrupt enable) bit in Register B is set.

  An interrupt can also be generated whenever the timekeeping update process is complete. This is a useful way to notify the MPU that a read of time or calendar information can now be done safely. This bit (B4 - UF), update-ended interrupt flag, is set to a one level whenever the update process ends. The interrupt will not be generated, however, unless the UIE enable bit in Register B is set.

  All of the interrupt flags in this register are cleared by a $\overline{\text{RESET}}$ or a reading of the register. The register must be read by the MPU'S program since the interrupt request line ($\overline{\text{IRQ}}$) does not specify which type of interrupt is requesting service. Once the Register is read by the CPU software, the Register bits may be tested and a determination made as to which interrupt will be processed.

- *Register D.* A single bit (B7) in Register D is used for RTC functions. The remaining seven bits always will read zeros. B7 is the valid RAM and time bit (VRT) and is used in conjunction with the power sense (PS) input pin. The power sense pin is typically connected to the chip power supply through a resistor/capacitor (RC) network. The RC network will hold the power sense input low as power is applied to the chip. During this time the VTR bit will read low, indicating that information contained in the various RAM locations is not valid. When power-up is complete, a read to Register D sets the VTR bit, indicating that RAM contents are valid.

USING A BATTERY BACKUP

A Real-Time Clock can be effective only if timekeeping functions are allowed to continue. In the event of system power failure a battery backup insures proper chip operation. If we assume, in this case, that the MPU will not be usable, the only burden on the RTC will be maintaining the proper time and date. In other words there will be no reads or writes made to the chip. Consequently, the chip enable ($\overline{CE}$) pin will be in the inactive state. Under these conditions, the powered-down MPU will not inadvertently receive power from the RTC chip. To insure that this power-saving feature is not thwarted, a pull-up resistor should be connected to the $\overline{CE}$ pin to prevent the line from going low.

II.11

Programmable Keyboard/Display Interface

PART NUMBER	8279 8279-5
FUNCTION	PROGRAMMABLE KEYBOARD/DISPLAY INTERFACE
MANUFACTURER	INTEL
VOLTAGES	Vcc = +5 Vss = GND
PWR. DISS.	1 W
PACKAGE	40-pin DIP
TEMPERATURE	0°C → +70°C
FEATURES	PROGRAMMABLE FOR SCANNED OR STROBED KEYBOARDS 2-KEY LOCKOUT OR N-KEY ROLLOVER WITH DEBOUNCE PROGRAMMABLE SCAN CLOCK RATE INTERRUPT CAPABILITY, 8/16 CHARACTER DISPLAY SIMULTANEOUS KEYBOARD AND DISPLAY OPERATION
COMPATIBLE MICROPROCESSORS	8085AH 8085A 8088
FUNCTIONAL DESCRIPTION	THIS CHIP IS DESIGNED TO HANDLE KEYBOARD INPUT AND DISPLAY OUTPUT FUNCTIONS FOR VARIOUS KEYBOARD AND DISPLAY TECHNOLOGIES. MODES OF OPERATION FOR INPUT AND OUTPUT ARE EASILY PROGRAMMED FROM THE HOST CPU. ONCE PROGRAMMED, THE CHIP TAKES OVER KEYBOARD AND DISPLAY FUNCTIONS. THE 8279 AND 8279-5 DIFFER IN SUPPLY VOLTAGE TOLERANCE AND TIMING CHARACTERISTICS.

Figure II.11.1 8279 Pinout
Courtesy of Intel Corp.

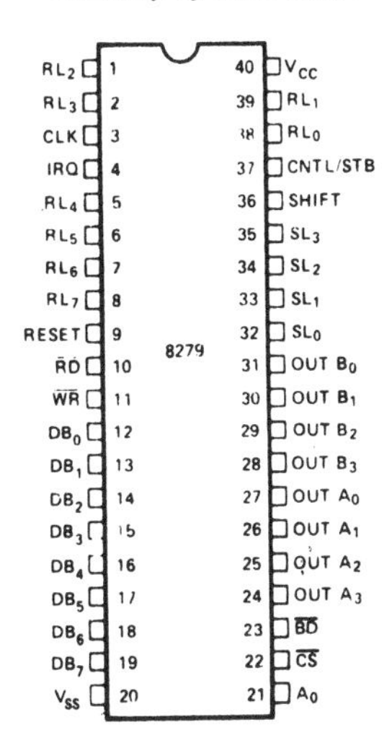

Figure II.11.2 Block Diagram
Courtesy of Intel Corp.

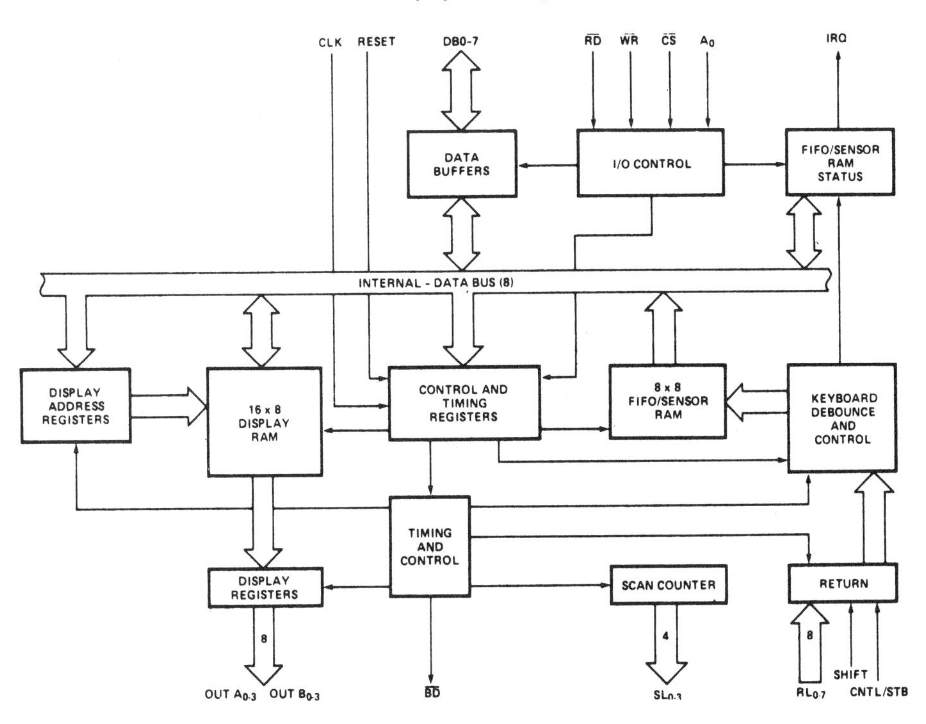

PIN NAME	PIN SYMBOL	PIN NUMBER	FUNCTION
RESET	RESET	9	A positive level on this pin resets the 8279. Immediately after a reset the chip defaults into the following modes: sixteen 8-bit character display, left entry; encoded scan keyboard with 2-key lockout; clock prescaler set to 31.
CLOCK	CLK	3	This input uses the same clock that drives the CPU. The specified minimum clock frequencies are 2 MHz for the 8279 and 3.125 MHz for the 8279-5. The clock input frequency is divided down via a programming command to produce an internal clock cycle of 10 μsec (100 kHz).
DATA BUS 0 DATA BUS 1 DATA BUS 2 DATA BUS 3 DATA BUS 4 DATA BUS 5 DATA BUS 6 DATA BUS 7	DB0 DB1 DB2 DB3 DB4 DB5 DB6 DB7	12 13 14 15 16 17 18 19	All programming commands and data are exchanged between the 8279 and the CPU on these lines.
READ WRITE	$\overline{RD}$ $\overline{WR}$	10 11	When low (only one line can be low at a time) these two CPU control lines enable the 8279 to either receive data from or send data to the CPU. This exchange of data (or commands) takes place on the data bus.

PIN NAME	PIN SYMBOL	PIN NUMBER	FUNCTION
CHIP SELECT	$\overline{CS}$	22	Commands or data may be transferred between the CPU and the 8279 when chip select is low. Chip select is derived by decoding a part of the CPU's address bus. Chip select does not have to be active for keyboard/display scanning operation to take place.
BUFFER ADDRESS	A0	21	The 8279 differentiates between the information present on the data bus with this line. If A0 is high then the data bus information is considered to be a Command or Status signal. If A0 is low, then the signals are considered to be data. A0 is typically one bit from the CPU's address bus, although not necessarily address bit A0.
SCAN LINE 0 SCAN LINE 1 SCAN LINE 2 SCAN LINE 3	SL0 SL1 SL2 SL3	32 33 34 35	These lines scan the keyboard switches and the display. Programming commands can set the scan lines to either decoded or encoded modes of operation.
RETURN LINE 0 RETURN LINE 1 RETURN LINE 2 RETURN LINE 3 RETURN LINE 4 RETURN LINE 5	RL0 RL1 RL2 RL3 RL4 RL5	38 39 1 2 5 6	Return lines are connected to the scan lines through sensors or keyboard switches. The return lines are internally pulled up to a high level. A switch clo-

(*continued*)

PIN NAME	PIN SYMBOL	PIN NUMBER	FUNCTION
RETURN LINE 6 RETURN LINE 7	RL6 RL7	7 8	sure brings a return line low.
SHIFT	SHIFT	36	This line is connected to the keyboard shift key. The line is internally pulled up to a high level. The line goes low when the shift key is pressed.
CONTROL/STROBE	CNTL/ STB	37	When used with keyboards, this line is connected to the control key. When the 8279 is set up in Strobed Input Mode, this line is connected to the strobe line. An internal pull-up resistor keeps this line high until a switch closure forces it low.
INTERRUPT REQUEST	IRQ	4	The active level for IRQ is high. An interrupt is requested when data is present in the keyboard FIFO. When data is read from the FIFO, the line goes low and returns high if FIFO data is still available. In Sensor mode, IRQ goes high whenever a change in a sensor level occurs.
OUTPUT A0 OUTPUT A1 OUTPUT A2 OUTPUT A3 OUTPUT B0 OUTPUT B1 OUTPUT B2	OUT A0 OUT A1 OUT A2 OUT A3 OUT B0 OUT B1 OUT B2	27 26 25 24 31 30 29	The output lines feed the displays as either two 4-bit ports or as one 8-bit port. The data on these lines is synchronized with the scan lines, allowing for multiplexed displays.

PIN NAME	PIN SYMBOL	PIN NUMBER	FUNCTION
OUTPUT B3	OUT B3	28	
BLANK DISPLAY	$\overline{\text{BD}}$	23	This line turns off the display. Line blanking occurs automatically or by command.
Vcc	Vcc	40	+5 V.
Vss	Vss	20	Ground.

Most microcomputer system designs require a keyboard or keypad as an input device. Many system designs also require a display, such as a seven-segment device, for system output. Consequently, a significant amount of hardware is allocated toward the support of these devices. Also, software must be written to handle the I/O activity for both the input and output functions. This software occupies memory space and, while executing, forces the CPU to suspend execution on other system software applications in order to handle keyboard input and display output operations.

The 8279 Programmable Keyboard/Display Interface (Figs. II.11.1 and II.11.2) is a powerful solution to both the hardware and software complexities of keypad and display interfacing. With this chip, standard keypads, keyboards, or arrays of switches/sensors may be controlled in a wide variety of modes. Simultaneously, a standard type of display (i.e., LED, incandescent, seven segment) can be scanned and updated. This concurrent keyboard/display activity takes place within the 8279 with a minimum of attention from the CPU, thus freeing the CPU for more important computational tasks. The chip is designed to be compatible with several popular microprocessors, thereby eliminating much control circuitry between the 8279 and host CPU.

8279 BASIC OPERATION

Four lines from the CPU control the 8279. Chip select ($\overline{\text{CS}}$) is activated by the decoding circuitry attached to the CPU address bus. When chip select is active (low), the 8279 is enabled to receive/transmit data or instructions with the CPU. When chip select is high, the chip is disabled from its sending or receiving operations with the CPU. However, the internal activity of the 8279 is not suspended. This means that keyboard and display scanning, keyboard input, and display output still take place.

The read and write ($\overline{RD}$, $\overline{WR}$) control lines come into play when the chip is selected. These lines determine whether the data on the bi-directional data bus (DB0-DB7) between the 8279 and the CPU is to be read (data from the 8279 to the CPU) or written (data from the CPU to the 8279). Read and write operations are always with respect to the CPU. That is, the CPU, not the peripheral chip, determines when reads and writes will occur.

When information is exchanged on the data bus, it is either in the form of keyboard/display data or 8279 programming commands. Control line A0 distinguishes between data and commands. If A0 is high when the chip is selected, then the information on the data bus is treated as command/status information. If A0 is low, then the information on the data bus is data. In both cases, $\overline{RD}$ or $\overline{WR}$ determine the direction of the information flow. A0 is typically one of the CPU's address bus lines (but not necessarily line A0). This means that a memory reference instruction in software distinguishes between data and command information by referencing two different addresses. For instance, assume that CPU address bit A0 is going to be used as the A0 line on the 8279. Further assume that the address DDFE (HEX) is allocated as the address for the 8279's data information. CPU bit A0 is "0" in this case (Fig. II.11.3). In this example, address DDFF (the next sequential address) would make CPU address bit A0 a "1," meaning that the information on the data bus is a command for the 8279 or status information from the 8279. To write a programming command to the 8279 using these addresses, a progamming instruction such as STA DDFF could be used. The implication here is that the accumulator register in the CPU was previously loaded with the desired command code. In a similar fashion, data could be read from the 8279 with an instruction such as LDA DDFE.

As mentioned earlier, CPU address bit A0 does not necessarily have to be the line connected to A0 on the 8279. Intel's SDK-85 kit implementation for this chip uses address 1800 for data and address 1900 for commands/status. This means that address bit A8 from the CPU is connected to A0 on the 8279. This simplifies the chip select decoding circuitry, but wastes addressing space. Using A8 as the control bit in this case allocates 512 memory locations for this operation when only two are needed. (Fig. II.11.4).

Figure II.11.3 How Pin A0 Distinguishes Between Data and Command Information

A_{15}	A_{14}	A_{13}	A_{12}	A_{11}	A_{10}	A_9	A_8	A_7	A_6	A_5	A_4	A_3	A_2	A_1	A_0	← Address Bus
1	1	0	1	1	1	0	1	1	1	1	1	1	1	1	0	← DDFE
1	1	0	1	1	1	0	1	1	1	1	1	1	1	1	1	← DDFF

Bit used as input to Line A0 (→ A_0)

Figure II.11.4 SKD-85 Kit Example

A_{15}	A_{14}	A_{13}	A_{12}	A_{11}	A_{10}	A_9	A_8	A_7	A_6	A_5	A_4	A_3	A_2	A_1	A_0	← Address Bus
0	0	0	1	1	0	0	0	0	0	0	0	0	0	0	0	← 1800
0	0	0	1	1	0	0	1	0	0	0	0	0	0	0	0	← 1900

This bit used as input to line A0 (↑ A_8)

PROGRAMMING THE 8279—KEYBOARD/DISPLAY INITIALIZATION

Eight op codes (Fig. II.11.5a) program the 8279 for any one of a number of possible configurations. Figure II.11.5b illustrates the general command structure for the eight op codes while Figure II.11.5c shows the structure for op code 000, Keyboard/Display Mode Set. This particular instruction uses two bits of encoded information to set up the display modes and three encoded bits to handle the various possibilities for keyboard usage. For instance, when the three least significant bits of this instruction are set to 000, the 8279 will be set to handle an encoded scan keyboard with 2-Key Lockout. These three bits allow several kinds of input devices to be handled by the 8279. The mode se-

Figure II.11.5 8279 Commands

Courtesy of Intel Corp.

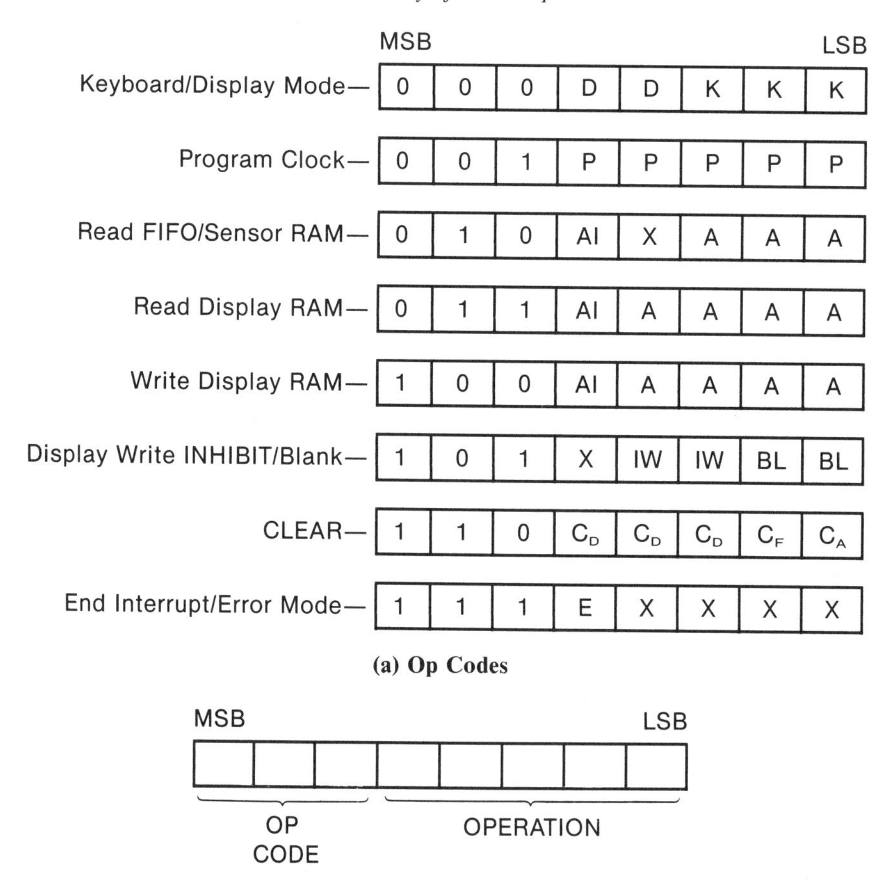

(a) Op Codes

(b) Command Format

Figure II.11.5 8279 Commands
(cont.)

Keyboard/Display Mode Set

	MSB							LSB
Code:	0	0	0	D	D	K	K	K

Where DD is the Display Mode and KKK is the Keyboard Mode.

DD		
0	0	8 8-bit character display — Left entry
0	1	16 8-bit character display — Left entry*
1	0	8 8-bit character display — Right entry
1	1	16 8-bit character display — Right entry

For description of right and left entry, see Interface Considerations. Note that when decoded scan is set in keyboard mode, the display is reduced to 4 characters independent of display mode set.

KKK			
0	0	0	Encoded Scan Keyboard — 2-Key Lockout*
0	0	1	Decoded Scan Keyboard — 2-Key Lockout
0	1	0	Encoded Scan Keyboard — N-Key Rollover
0	1	1	Decoded Scan Keyboard — N-Key Rollover
1	0	0	Encoded Scan Sensor Matrix
1	0	1	Decoded Scan Sensor Matrix
1	1	0	Strobed Input, Encoded Display Scan
1	1	1	Strobed Input, Decoded Display Scan

(c) Keyboard/Display Mode Set Command Details

lected determines how the chip responds to an input device. Most input devices consist of a series of switches that either open or close when struck. The chip is not limited to switches as input devices since it is possible to connect sensors to the chip as well.

Most input options are for scanned keyboards which depend on the chip to generate keyboard scanning signals (SL0-SL3). When a key is struck, a portion of the scan signal is returned to the 8279 on Return Lines RL0-RL7 and stored in a FIFO RAM. The initial scanning information and the return information determine which key was struck. The scanning signal can be programmed to be either in encoded (8 × 8 keyboard) or decoded (4 × 8 keyboard) form. In an encoded form some additional decoding circuitry between the 8279 and the keyboard is necessary. A 74LS156 or similar device can be used for this decoding. The 8279 also responds to strobed keyboards by transferring the key information to the FIFO RAM upon activation of a strobe signal (rising edge on pin 37, CNTL/STB). This input pin is connected to the Control key for scanned operation.

For most keyboarding operations, when a key is struck, the 8279 first detects a key closure and then waits 10 msec to allow for key debounce. If the key is still closed after this time, then the scan, return, Shift, and Control line information is transferred to the FIFO RAM. The CPU is notified that a key was pressed via the Interrupt Request line. The FIFO RAM is eight bytes deep and will generate an interrupt whenever it contains data. The CPU can then execute software to collect and process the FIFO data. A status register within the 8279 keeps track of the FIFO and monitors error conditions that may occur (Fig. II.11.6).

For encoded or decoded scan Sensor Matrix mode, switch information is obtained in a different fashion. This mode uses an array of switches or sensors as the input device. Each switch is mapped to one of the FIFO RAM locations. That is, an 8 × 8 array of switches can be mapped directly to each bit in the FIFO RAM. When a key position is changed, the location in FIFO RAM that represents the switch also is changed. An interrupt notifies the CPU of the change. The debouncing circuitry is disabled in this mode.

Two other options exist for scanned keyboard operation: 2-Key Lockout or N-Key Rollover. With 2-Key Lockout, only one key depression is accepted and eventually entered into the FIFO RAM. If N-Key Rollover is selected then more than one key depression can be detected and entered into the FIFO. If two or more keys are depressed simultaneously, then the keys are entered into the FIFO RAM in the order in which they were found by the keyboard scanning circuitry.

There are two bits in the Keyboard/Display Mode Set op code that control the display mode. They allow either 8- or 16-character (eight bits each) display operations with left or right data entry. An exception to these modes occurs if decoded scan is selected for keyboard operation. In this case the display will only be 4 characters long.

Figure II.11.6 FIFO Status Word
Courtesy of Intel Corp.

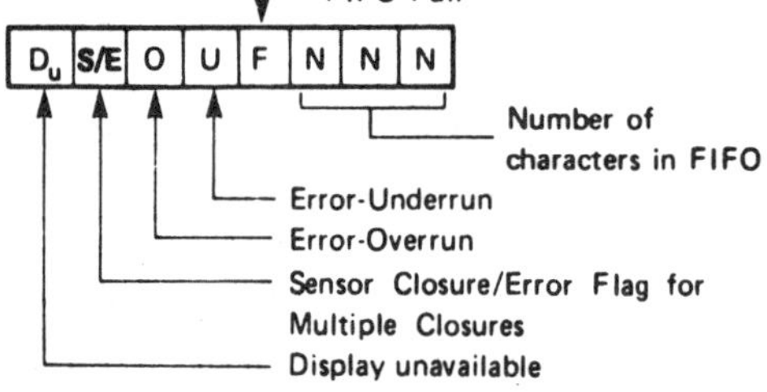

FIFO Full

D_u	S/E	O	U	F	N	N	N

Number of characters in FIFO
Error-Underrun
Error-Overrun
Sensor Closure/Error Flag for Multiple Closures
Display unavailable

Data for display is stored in a 16 × 8 Display RAM. This RAM is organized to be read as a byte-wide output or as two 4-bit nibbles. Data in the RAM is first transferred to a Display Register and then to output pins OUT A0-A3 and OUT B0-B3. Data is placed into the RAM from the CPU. The CPU may also read the RAM information. A Display Address Register keeps track of the address of the data in the RAM. The address is placed in the address register by a programming command from the CPU. The programming command allows any address to be specified and allows auto increment operation. This means that after each read or write to the Display RAM, the Display Address Register will automatically be updated to point to the next sequential address.

Right entry means that as data is transferred from the Display RAM to the display, each succeeding entry appears in the right-most character space (the display is shifted left with each new entry). If the display were operated slowly enough, data characters could be seen entering from the right and working their way left through the display. The first character entered would eventually "fall off" the left end of the display and be lost. Left entry is similar except that data enters the display from the left side. If the auto increment option is used, the next Display RAM location provides data for the next character display device after every character transfer. This means that the Display RAM contents are being read out to the display in a sequential fashion. If auto increment is not used, then the same Display RAM location specified provides data only to its corresponding display device. In other words, only one of the 8 or 16 possible display devices will be activated.

Programming the 8279 Clock

The typical times specified for 8279 operation are derived from the chip clock input. Keyboard scan time is given as 5.1 msec, keyboard debounce time is 10.3 msec, and blanking time is 160 μsec. Many other time specifications are listed in the data book but the question is, how are these times generated? Locking the 8279 chip to a specific input clock frequency would severely limit the use of this chip in many microprocessor systems. Fortunately, the programmability of the 8279 overcomes this limitation. The proper scan times can be obtained from almost any existing clock frequency by carefully programming the 8279 Clock Prescaler. Op code 001 is sent to the 8279 with five bits ranging in value from 2 to 31. The idea is to divide down the clock frequency presented to the 8279 (whatever it is) to obtain an internal frequency of 100 kHz (10 μsec). The clock input frequency is divided down by the 5-bit number specified in the program command. This frequency (100 kHz) is used to generate the proper internal timings for scan, debounce, etc. For instance, if the clock frequency presented to the 8279 was 1 MHz, then this frequency would have to be divided by ten (01010) to obtain an internal frequency of 100 kHz. The 8279 command, 00101010, sets the prescaler to 10.

OTHER 8279 FEATURES

Three remaining programming commands give the 8279 additional capabilities. The Display Write Inhibit/Blanking command allows masking of the output data to both blank the display and inhibit writing to the display RAM. The command allows a byte of output data to be treated as two 4-bit nibbles. An individual bit in the command is used to modify the A or B nibble. For instance, to blank the portion of the display corresponding to nibble B the bit in the command representing blanking for nibble B would be set to one, in which case the display would be turned off. All of these control bits are zeroed upon a reset.

The Clear command allows the display RAM to be cleared to a known state. The command will, when enabled, clear all the rows of the display RAM to zeros, ones, or Hex 20. This operation takes approximately 160 μsec. The FIFO status word's (Fig. II.11.6) most significant bit is set to one under these circumstances and is reset to one when the clear operation is complete. Furthermore, the Clear command can clear the status word, reset the interrupt output line, and reposition the sensor RAM row pointer to row zero.

The End Interrupt/Error Mode Set command has two uses. When used with a switch matrix (Sensor Matrix Mode), it brings the IRQ line low. For example, assume that an interrupt occurred because a sensor switch changed level. Normally the 8279 inhibits further writing into the sensor RAM until the interrupt is processed. This command overrides this normal sequence of events by inhibiting the interrupt line from becoming active. This command, through the "E" bit, can also program a special error mode for N-key rollover operation. When the "E" bit is set, any multiple key closure detected during a single debounce cycle sets the Sensor Closure/Error Flag in the FIFO status word, raises the interrupt line, and inhibits further writing to the FIFO RAM.

Finally, the FIFO status word also indicates other conditions occurring within the FIFO RAM. If the FIFO RAM is full and another character is written to the RAM, an overrun error occurs and an overrun flag is set in the status word. Similarly, an underrun flag is set when the CPU attempts to read an empty FIFO RAM. Another flag in the status word indicates that the FIFO RAM is full, while three additional bits in the status word indicate the number of characters in the FIFO RAM.

II.12

CRT Controller

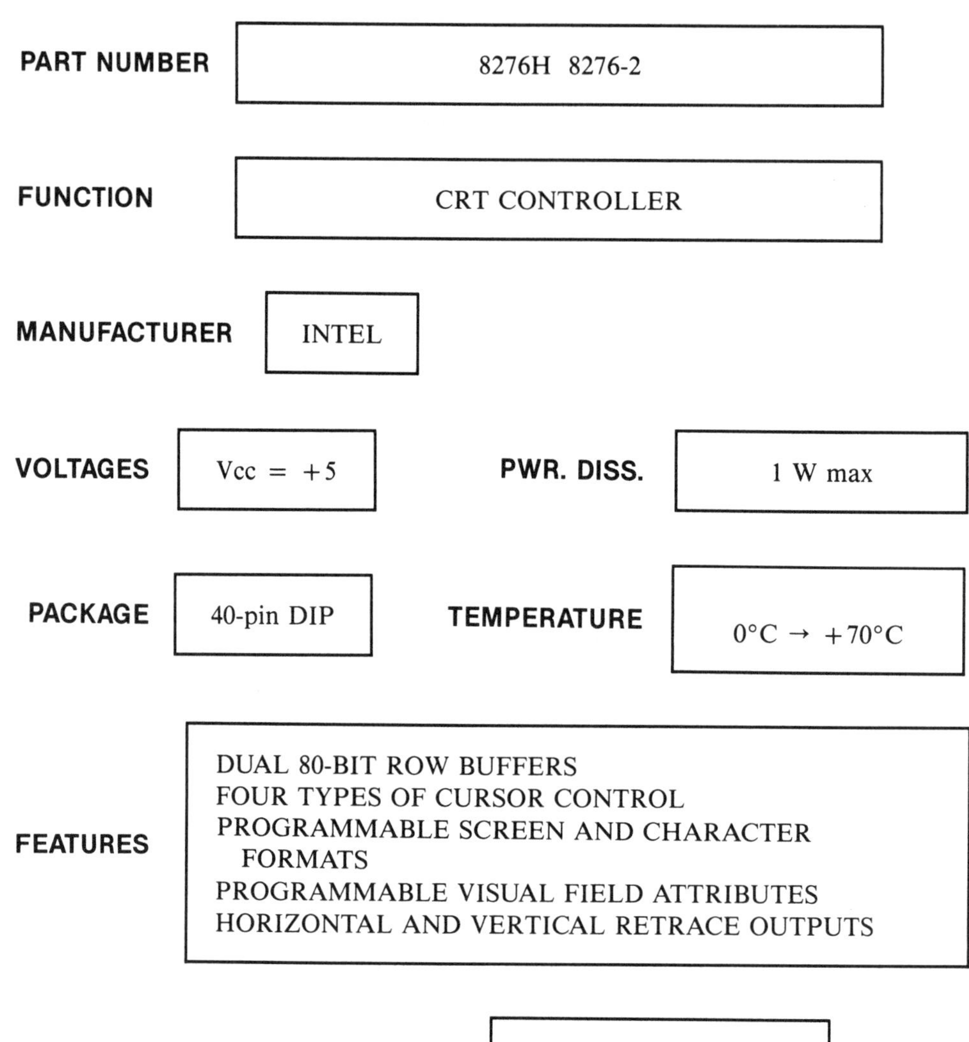

PART NUMBER: 8276H 8276-2

FUNCTION: CRT CONTROLLER

MANUFACTURER: INTEL

VOLTAGES: Vcc = +5

PWR. DISS.: 1 W max

PACKAGE: 40-pin DIP

TEMPERATURE: 0°C → +70°C

FEATURES:

DUAL 80-BIT ROW BUFFERS
FOUR TYPES OF CURSOR CONTROL
PROGRAMMABLE SCREEN AND CHARACTER FORMATS
PROGRAMMABLE VISUAL FIELD ATTRIBUTES
HORIZONTAL AND VERTICAL RETRACE OUTPUTS

COMPATIBLE MICROPROCESSORS: 8051 8085 8086 8088

FUNCTIONAL DESCRIPTION:

THE CRT CONTROLLER FORMS AN INTERFACE BETWEEN A RASTER SCAN CRT AND MICROPROCESSOR. THE CONTROLLER ACCEPTS DATA FROM MEMORY AND BUFFERS IT INTERNALLY FOR EVENTUAL CRT DISPLAY. VIDEO CONTROL SIGNALS ARE GENERATED TO REGULATE CRT HORIZONTAL AND VERTICAL FUNCTIONS. PROGRAMMABLE ATTRIBUTES ALLOW DISPLAY HIGHLIGHT, REVERSE VIDEO, UNDERLINE, AND BLINK.

Figure II.12.1 8276 Block Diagram

Courtesy of Intel Corp.

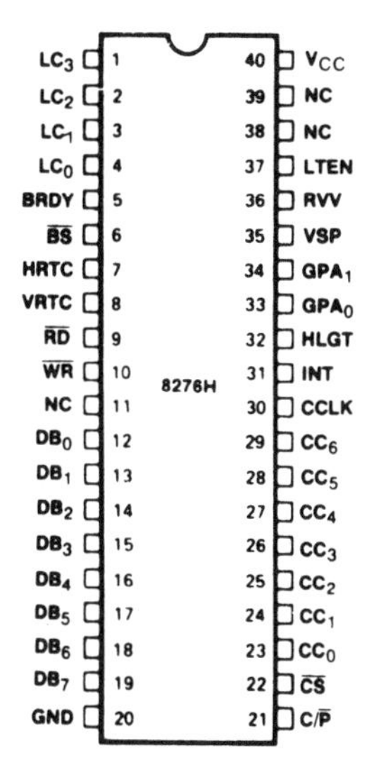

Figure II.12.2 8276 Pinout

Courtesy of Intel Corp.

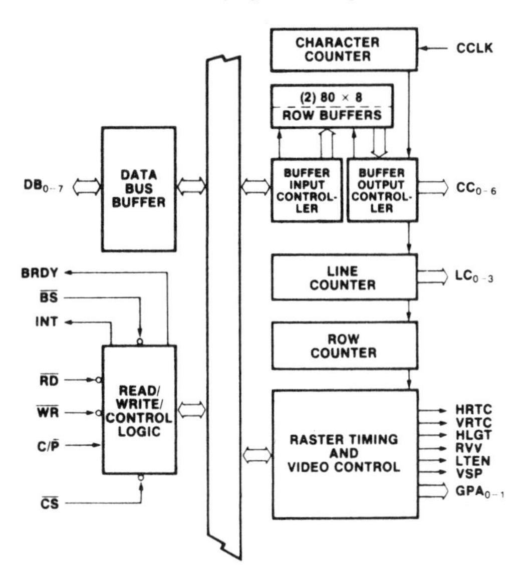

PIN NAME	PIN SYMBOL	PIN NUMBER	FUNCTION
CHIP SELECT	$\overline{CS}$	22	$\overline{CS}$, when low, enables the 8276 for read and write operations.
READ INPUT	$\overline{RD}$	9	Provided that $\overline{CS}$ is active, a low level on RD, in conjunction with $C/\overline{P} = 1$, places the contents of the Status Register onto the Data Bus.
WRITE INPUT	$\overline{WR}$	10	When low, along with an active $\overline{CS}$, the 8276 is enabled to receive data or control information from the CPU.
BUFFER SELECT	$\overline{BS}$	6	A low level on $\overline{BS}$ enables data to be written into the chip Row Buffers.
BUFFER READY	BRDY	5	This output line, when high, indicates that the 8276 is ready to receive character data.
PORT ADDRESS	$C/\overline{P}$	21	The state on this input pin determines whether the chip will engage in a read or write of Status, Parameter, or Command information.
INTERRUPT OUTPUT	INT	31	A high level on this output is used to signal the CPU that an interrupt request is in process.
DATA BUS 0 DATA BUS 1	DB0 DB1	12 13	The data bus connects the CPU to the CRT Control

PIN NAME	PIN SYMBOL	PIN NUMBER	FUNCTION
DATA BUS 2 DATA BUS 3 DATA BUS 4 DATA BUS 5 DATA BUS 6 DATA BUS 7	DB2 DB3 DB4 DB5 DB6 DB7	14 15 16 17 18 19	chip. Control information and data are passed along the bidirectional, tri-state data bus.
CHARACTER CODE 0 CHARACTER CODE 1 CHARACTER CODE 2 CHARACTER CODE 3 CHARACTER CODE 4 CHARACTER CODE 5 CHARACTER CODE 6	CC0 CC1 CC2 CC3 CC4 CC5 CC6	23 24 25 26 27 28 29	The character codes that determine what will be displayed on the CRT, in conjunction with a ROM chip, are output here. These outputs are designed to drive an external Character Generator.
LINE COUNT 0 LINE COUNT 1 LINE COUNT 2 LINE COUNT 3	LC0 LC1 LC2 LC3	4 3 2 1	The line count is also used with the character codes to determine the CRT display.
CHARACTER CLOCK	CCLK	30	This clock input, which is related to the CRT's display timing requirements, is obtained from external Dot-Clock circuitry.
HORIZONTAL RETRACE	HRTC	7	This output, which is programmable, is used to control the horizontal retrace time. VSP is high and LTEN is low during this time.

(*continued*)

PIN NAME	PIN SYMBOL	PIN NUMBER	FUNCTION
VERTICAL RETRACE	VRTC	8	This programmable output is used to control the vertical retrace time. VSP is high and LTEN is low during this time.
VIDEO SUPPRESSION	VSP	35	A high level on this output is used to blank the video signal. Many conditions, such as retrace intervals, can activate this signal.
LIGHT ENABLE	LTEN	37	Light Enable is active high under program control. It is used to support the chip's underlining feature.
REVERSE VIDEO	RVV	36	The CRT video display can be reversed by activating this signal.
HIGHLIGHT	HLGT	32	Field attribute codes can specify when CRT display information is to be intensified. This output signal controls that function.
GENERAL PURPOSE ATTRIBUTE CODES	GPA0 GPA1	33 34	These two optional outputs can be activated through the use of the general purpose field attribute codes.
+5 V POWER SUPPLY	Vcc	40	+5 V.
GROUND	GND	20	Ground.
NO CONNECTION	NC	11	No connection to these pins is required.
NO CONNECTION	NC	38	
NO CONNECTION	NC	39	

A usable CRT monitor display is obtained through a complex process that involves the conversion of digital information into a video information stream. In addition to the conversion of parallel digital codes into serial analog video signals, this process also requires circuitry for the control of character timing, horizontal retrace, and vertical retrace. To prevent screen flicker, data must be supplied to the CRT at a reasonable rate and must also be supplied at a rate which will not bog down a controlling microprocessor through inefficient system bus utilization. These fundamental design objectives, require a substantial amount of control circuitry. These objectives, including a host of other useful display enhancements, are best transformed into a practical display system through the use of a CRT Controller.

OVERVIEW OF A CRT DISPLAY SYSTEM

The CRT Controller will not in itself provide all the capabilities necessary for a functional display system. A considerable amount of external circuitry is still required to drive the CRT. The controller chip will ease the microprocessor/CRT interface design work through its programmability and CPU interface support logic.

In a typical, non-graphics display terminal, a memory area is set aside to store data for display. A controlling microprocessor is called upon to retrieve this data and supply it to the CRT circuitry. There are many possible ways to implement this process with each involving a trade off between CPU data retrieval overhead and complex controlling circuitry. The 8276 (Figs. II.12.1 and II.12.2) simplifies this process with two, 80-character Row Buffers that can keep a supply of data available for display. The CPU periodically fills the buffers, but between fill operations is free to do other useful work. The use of Row Buffers is purported to reduce CPU and system bus usage to a maximum of 25% as opposed to the 90–95% usage that would be required if the CPU supplied data, byte by byte, directly to the CRT.

The data obtained from system memory can be in any form suitable for the application at hand. Most common terminals and processing systems use the ASCII code for alphanumeric display. Regardless of the code used, it must be converted to match the needs of the CRT display device. The data information for display is stored in system memory in parallel groups of bits. In contrast, the CRT electron beam sends information to the screen in a serial fashion. One function of the controller and support circuitry is to convert the parallel data information into a serial video stream. Additionally, the data from memory must be changed into character data that will be visually meaningful on the CRT screen. For example, the ASCII code for the letter A is 1000001. The displayed version of the letter A is a matrix of dots which, when illuminated by the electron beam, take on the shape of the letter. The controller aids the process of converting meaningful CPU information to CRT display information by automatically sending out the Row buffered data on Character Code lines (CC0-CC6) and Line Count outputs (LC0-LC3). These lines are designed to drive a Character Generator PROM that will, in turn, provide the proper bit signals (dots) for the CRT. A final parallel-to-serial conversion takes place before the dot information is sent on to the CRT electronics.

Timing information must also be provided to properly sequence the character and ultimate CRT dot information. A Dot Clock formed from a crystal oscillator generates the basic timing frequency. The frequency for this clock depends on the CRT chosen, the number of characters per display line, the number of display lines per screen, and other factors. The Character Clock (CCLK) input to the CRT Controller chip, also based on these parameters, is derived from the Dot Clock.

The CRT Controller also aids in the development of horizontal and vertical timing pulses. These pulses control the retrace time, which occurs under two conditions. After a line of data is displayed, the electron beam must sweep from the right side of the screen back to the left (Horizontal retrace). After a screen (frame) of information has been displayed, the beam sweeps from bottom to top (Vertical retrace). The controller provides a video suppression signal (VSP), as well, because during retrace, the electron beam must be turned off (blanked). The designer can adjust display height and width by controlling retrace times. Using the controller, synchronization between the video, horizontal, vertical, and suppression signals is maintained. Figure II.12.3 illustrates the basic circuitry for a CRT Controller-based display system.

Interfacing the 8276 to a Microprocessor

The host microprocessor in a terminal display system is generally dedicated to sending data to the CRT Controller for display and maintaining data in a display memory area. The display memory contains data that eventually will be displayed on the CRT screen. Since the data frequently changes (keyboard entries, program output, etc.), the microprocessor divides its time between memory and controller duties.

Figure II.12.3 Basic CRT Controller Circuitry

Courtesy of Intel Corp.

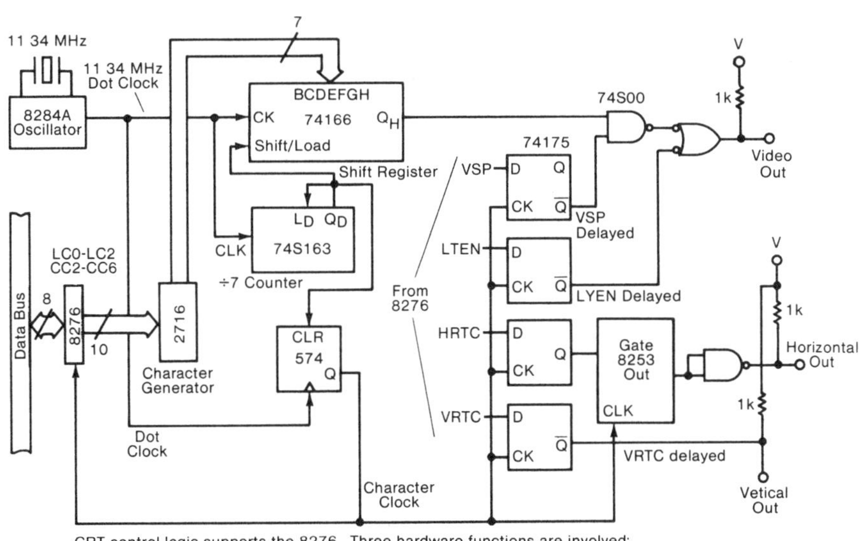

CRT control logic supports the 8276. Three hardware functions are involved: a dot/character clock oscillator, an EPROM character generator, and a character-shift register.

Therefore, minimizing CPU involvement in character data transfer can enhance system performance.

The CPU communicates with the CRT Controller over the data bus. The actual information exchanged can run the gamut from display data to controller commands. Several control signals regulate information transfer. Standard signals such as Read ($\overline{RD}$), Write ($\overline{WR}$), and Chip Select ($\overline{CS}$) control the input and output of commands, status, and parameter information. At the appropriate time in a transfer cycle these active low signals are asserted either directly by the microprocessor ($\overline{RD}$ and $\overline{WR}$) or indirectly ($\overline{CS}$) through CPU address bus decoding logic.

The C/$\overline{P}$ (Port Address) input line, distinguishes between the various sources and destinations of data bus information. Figure II.12.4 shows the meaning of the C/$\overline{P}$ input. For instance, a write operation performed under the direction of $\overline{CS}$ = 0 and C/$\overline{P}$ = 1 directs the CPU Data Bus information to the 8276 Command Register.

The CRT Controller can accept CPU information without Chip Select being active. Two additional lines, $\overline{BS}$ (Buffer Select) and BRDY (Buffer Ready), are used to route data to the 8276 Row Buffers for display. By bringing BRDY high, the 8276 informs the CPU that it is

Figure II.12.4 C/$\overline{P}$ Options

Courtesy of Intel Corp.

C/$\overline{P}$	OPERATION	REGISTER
0	Read	RESERVED
0	Write	PARAMETER
1	Read	STATUS
1	Write	COMMAND

ready to accept display data. This output often is used as a CPU interrupting signal. The CPU lowers $\overline{BS}$ and then $\overline{WR}$ when display data destined for the Row Buffers is placed on the data bus. Buffer Select is usually generated as a unique address bus decode in a manner similar to the generation of $\overline{CS}$ so that several addresses in a memory-mapped scheme, or several ports in an I/O scheme form the communication link between CPU and controller. There are two Row Buffers within the CRT Controller that are automatically switched so that one outputs data for display while the second is filling with data from the CPU. Figure II.12.5 summarizes the functions of the control lines. A separate interrupt line (INT) is also provided that can be programmed to generate an interrupt signal at the end of each frame if desired. This may be useful for periodic inquiries of controller status.

Interfacing the 8276 to a CRT

Once the Row Buffers contain character data, the controller outputs this information at a rate determined by the Character Clock (CCLK). The Character Clock frequency is obtained by dividing the external Dot Clock frequency by the number of dots used to create the width of the displayed character (CRT characters are displayed as

Figure II.12.5 Control Line Functions

Courtesy of Intel Corp.

C/$\overline{P}$	$\overline{RD}$	$\overline{WR}$	$\overline{CS}$	$\overline{BS}$	
0	0	1	0	1	Reserved
0	1	0	0	1	Write 8276H Parameter
1	0	1	0	1	Read 8276H Status
1	1	0	0	1	Write 8276H Command
X	1	0	1	0	Write 8276H Row Buffer
X	1	1	X	X	High Impedance
X	X	X	1	1	High Impedance

groupings of individual dots). Both the Dot Clock and number of dots per character are application dependent.

The character data output by the controller is, by design, intended to feed a Character Generator. Character Generators can be purchased as ROM packages or custom made by programming a PROM/EPROM. The 8276 accesses Character Generator locations, which in turn provide the proper dot outputs for creation of the CRT display figure. The CRT Controller has two sets of output lines designated for character data output. Output lines CC0-CC6 (Character Codes) and LC0-LC3 (Line Count) feed directly to the Character Generator address inputs. Each character displayed on the CRT actually consists of several rows of dots which constitute the entire character. For each value of Line Count (programmable), the Row Buffer character data information will be output on the Character Code lines. After one horizontal scan across the CRT, the Line Count value is incremented by one, and then the same information in the Row Buffer is read out again. This procedure will continue until the entire character is formed. The Line Count value is then restarted and the second Row Buffer begins to output its character data. The process just described repeats and keeps data continuously flowing, as video information to the screen. Figure II.12.6 details an example of this process.

In addition to display information, the CRT Controller also produces useful CRT synchronization signals. Horizontal and vertical retrace pulses are generated by the 8276 and are properly synchronzied with the character information. The duration of these pulses is programmable. The horizontal retrace signal can be set between 2 and 32 CCLK periods. The vertical retrace signal can be set between 1 and 4 row time intervals. This allows timing flexibility so the controller can interface to a variety of CRTs. In order to prevent a trace from appearing on the screen during the retrace periods, a video suppression output (VSP) is used to inhibit the video signal. Several other video signals are included to control special features.

UNDERSTANDING THE 8276 REGISTERS AND COUNTERS

A number of registers and counters set the CRT Controller chip operating characteristics. A read-only Status Register (Fig. II.12.7) can be accessed over the data bus at any time. The individual bits in this register monitor chip operations.

The Command Register and Parameter Register work in tandem

Figure II.12.6 Display Example

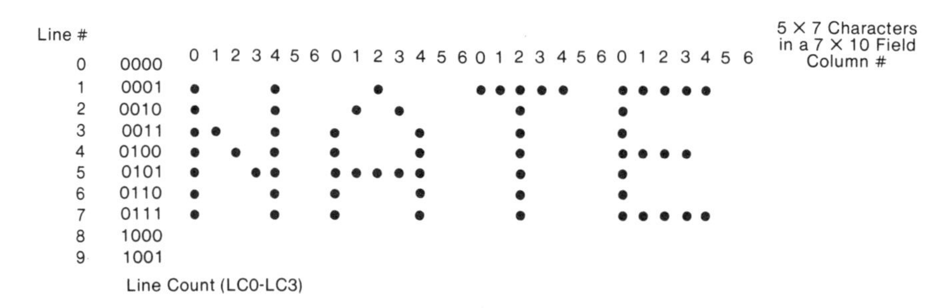

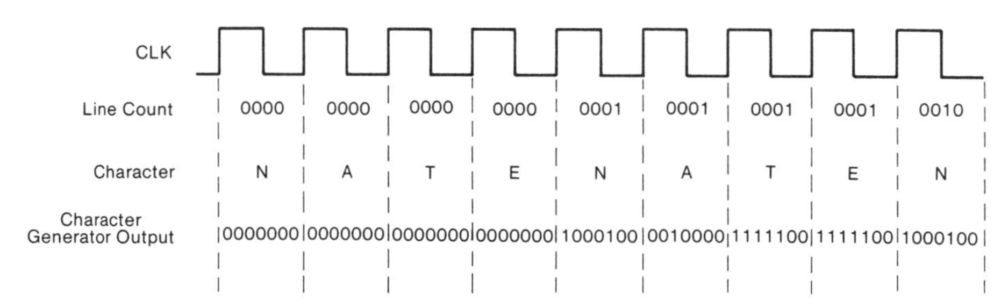

to set the chip for a particular operating procedure. The CPU initializes the controller chip for display characteristics such as characters per row, rows per frame, and underline placement, with commands and appropriate parameter information. Figure II.12.8 lays out the command and parameter definition for this chip. The amount of parameter information sent with a command varies with the command chosen. A byte of information on the data bus finds its way to the Command Register during a write operation when the $C/\overline{P}$ bit is high. The command chosen (there are seven) may or may not require additional parameter information. If required, subsequent writes with $C/\overline{P}$ low places parameter information into the Parameter Register. For example, the Reset Command stipulates that four parameter bytes are required. These parameters in particular will set many of the counters which control display format. The Start Display Command enables the signals that control Row Buffer fill and video output operations.

A programmable Character Counter sets the number of characters that are displayed per row. For instance, 80 characters per row is com-

Figure II.12.7 Status Flags

Courtesy of Intel Corp.

	OPERATION	C/P	DESCRIPTION	DATA BUS MSB LSB
Command	Read	1	Status Word	0 IE IR X IC VE BU X

IE — (Interrupt Enable) Set or reset by command. It enables vertical retrace interrupt. It is automatically set by a "Start Display" command and reset with the "Reset" command.

IR — (Interrupt Request) This flag is set at the beginning of display of the last row of the frame if the interrupt enable flag is set. It is reset after a status read operation.

IC — (Improper Command) This flag is set when a command parameter string is too long or too short. The flag is automatically reset after a status read.

VE — (Video Enable) This flag indicates that video operation of the CRT is enabled. This flag is set on a "Start Display" command, and reset on a "Stop Display" or "Reset" command.

BU — (Buffer Underrun) This flag is set whenever a Row Buffer is not filled with character data in time for a buffer swap required by the display. Upon activation of this bit, buffer loading ceases, and the screen is blanked until after the vertical retrace interval.

mon in many computer display terminals and this value can be set using the first parameter byte of the Reset Command. Any number of characters between 1 and 80 is valid.

A Row Counter, also programmable with the Reset Command, specifies the number of rows per frame. This value can vary between 1 and 64 rows.

The Line Counter determines character height. Several horizontal scans will be required to completely create a character on the screen. The Line Counter, whose contents are output at LC0-LC3, can be set between 1 and 16 with the Reset Command.

The numerical values loaded initially into these counters allow the display system designer considerable versatility when setting up the "looks" of the output display. Furthermore, as mentioned earlier, the horizontal and vertical retrace times are related to these counts.

Figure II.12.8 Instruction Set

Courtesy of Intel Corp.

The 8276H instruction set consists of 7 commands.

COMMAND	NO. OF PARAMETER BYTES
Reset	4
Start Display	0
Stop Display	0
Load Cursor	2
Enable Interrupt	0
Disable Interrupt	0
Preset Counters	0

In addition, the status of the 8276H can be read by the CPU at any time.

1. RESET COMMAND

	OPERATION	C/P	DESCRIPTION	DATA BUS MSB … LSB
Command	Write	1	Reset Command	0 0 0 0 0 0 0 0
Parameters	Write	0	Screen Comp Byte 1	S H H H H H H H
	Write	0	Screen Comp Byte 2	V V R R R R R R
	Write	0	Screen Comp Byte 3	U U U U L L L L
	Write	0	Screen Comp Byte 4	M 1 C C Z Z Z Z

Action—After the reset command is written, BRDY goes inactive, 8276H interrupts are disabled, and the VSP output is used to blank the screen. HRTC and VRTC continue to run. HRTC and VRTC timing are random on power-up.

As parameters are written, the screen composition is defined.

Parameter—S Spaced Rows

S	FUNCTIONS
0	Normal Rows
1	Spaced Rows

Parameter—HHHHHHH Horizontal Characters/Row

H H H H H H H	NO. OF CHARACTERS PER ROW
0 0 0 0 0 0 0	1
0 0 0 0 0 0 1	2
0 0 0 0 0 1 0	3
.	.
.	.
.	.
1 0 0 1 1 1 1	80
1 0 1 0 0 0 0	Undefined
.	.
.	.
.	.
1 1 1 1 1 1 1	Undefined

Parameter—VV Vertical Retrace Row Count

V V	NO. OF ROW COUNTS PER VRTC
0 0	1
0 1	2
1 0	3
1 1	4

Parameter—RRRRRR Vertical Rows/Frame

R R R R R R	NO. OF ROWS/FRAME
0 0 0 0 0 0	1
0 0 0 0 0 1	2
0 0 0 0 1 0	3
.	.
.	.
.	.
1 1 1 1 1 1	64

Parameter—UUUU Underline Placement

U U U U	LINE NUMBER OF UNDERLINE
0 0 0 0	1
0 0 0 1	2
0 0 1 0	3
.	.
.	.
.	.
1 1 1 1	16

Parameter—LLLL Number of Lines per Character Row

L L L L	NO. OF LINES/ROW
0 0 0 0	1
0 0 0 1	2
0 0 1 0	3
.	.
.	.
.	.
1 1 1 1	16

Parameter—M Line Counter Mode

M	LINE COUNTER MODE
0	Mode 0 (Non-Offset)
1	Mode 1 (Offset by 1 Count)

Parameter—CC Cursor Format

C C	CURSOR FORMAT
0 0	Blinking reverse video block
0 1	Blinking underline
1 0	Non-blinking reverse video block
1 1	Non-blinking underline

Figure II.12.8 Instruction Set
(cont.)
Courtesy of Intel Corp.

Parameter—ZZZZ Horizontal Retrace Count

Z Z Z Z	NO. OF CHARACTER COUNTS PER HRTC
0 0 0 0	2
0 0 0 1	4
0 0 1 0	6
.	.
.	.
.	.
1 1 1 1	32

Note: uuuu MSB determines blanking of top and bottom lines (1 = blanked, 0 = not blanked).

2. START DISPLAY COMMAND

	OPERATION	C/P	DESCRIPTION	DATA BUS MSB LSB
Command	Write	1	Start Display	0 0 1 0 0 0 0 0
No parameters				

Action—8276H interrupts are enabled, BRDY goes active, video is enabled, Interrupt Enable and Video Enable status flags are set.

3. STOP DISPLAY COMMAND

	OPERATION	C/P	DESCRIPTION	DATA BUS MSB LSB
Command	Write	1	Stop Display	0 1 0 0 0 0 0 0
No parameters				

Action—Disables video, interrupts remain enabled, HRTC and VRTC continue to run, Video Enable status flag is reset, and the "Start Display" command must be given to reenable the display.

4. LOAD CURSOR POSITION

	OPERATION	C/P	DESCRIPTION	DATA BUS MSB LSB
Command	Write	1	Load Cursor	1 0 0 0 0 0 0 0
Parameters	Write Write	0 0	Char. Number Row Number	(Char. Position in Row) (Row Number)

Action—The 8276H is conditioned to place the next two parameter bytes into the cursor position registers. Status flag not affected.

5. ENABLE INTERRUPT COMMAND

	OPERATION	C/P	DESCRIPTION	DATA BUS MSB LSB
Command	Write	1	Enable Interrupt	1 0 1 0 0 0 0 0
No parameters				

Action—The interrupt enable flag is set and interrupts are enabled.

6. DISABLE INTERRUPT COMMAND

	OPERATION	C/P	DESCRIPTION	DATA BUS MSB LSB
Command	Write	1	Disable Interrupt	1 1 0 0 0 0 0 0
No parameters				

Action—Interrupts are disabled and the interrupt enable status flag is reset.

7. PRESET COUNTERS COMMAND

	OPERATION	C/P	DESCRIPTION	DATA BUS MSB LSB
Command	Write	1	Preset Counters	1 1 1 0 0 0 0 0
No parameters				

Action—The internal timing counters are preset, corresponding to a screen display position at the top left corner. Two character clocks are required for this operation. The counters will remain in this state until any other command is given.

This command is useful for system debug and synchronization of clustered CRT displays on a single CPU. After this command, two additional clock cycles are required before the first character of the first row is put out.

8276 SPECIAL FEATURES

Through the use of built-in code and attribute features, CPU software overhead can be further reduced as the visual look of the display is modified.

Special software codes can be sent by the CPU to notify the controller of End of Row and End of Screen conditions. When End of Row is received by the controller, the remainder of that row is blanked on the screen. End of Screen is similar in that the remainder of the screen is blanked when this code is received. End of Row-Stop Buffer Loading and End of Screen-Stop Buffer Loading are special code variations that have the same visual effects already mentioned. The codes with the Stop Buffer Loading feature minimize the amount of character data sent from the CPU to controller. The Stop Buffer Loading option means that no additional character bytes will be sent following the code. The lack of this option does not stop the flow of character data; the controller simply ignores it. For display systems with a specific visual format, the Stop Buffer Loading option can reduce display memory requirements.

Obviously the CRT Controller must be able to distinguish between character data and special codes since they both arrive on the same data bus. The data bus is eight bits wide, but only seven of these bits are sent on to the Character Generator. The most significant bit of the data bus, when zero, designates the byte as character data. When the most significant bit is one, the byte is interpreted as a special code or field attribute. Figure II.12.9 shows the byte arrangement for the special codes just discussed.

Six field attributes affect the display visual appearance. Figure II.12.10 illustrates the field attribute code. The individual bits, which may be used simultaneously, control the highlight, blink, reverse video, underline, and general purpose attributes. When the attribute code is sent to the controller, it affects all characters as far as either the end of the screen or the next field attribute code. Attribute codes are reset

Figure II.12.9 Special Control Character
Courtesy of Intel Corp.

SPECIAL CONTROL CHARACTER

MSB ... LSB
1 1 1 1 0 0 S S — SPECIAL CONTROL CODE

S	S	FUNCTION
0	0	End of Row
0	1	End of Row-Stop Buffer Loading
1	0	End of Screen
1	1	End of Screen-Stop Buffer Loading

Figure II.12.10 Field Attribute Code
Courtesy of Intel Corp.

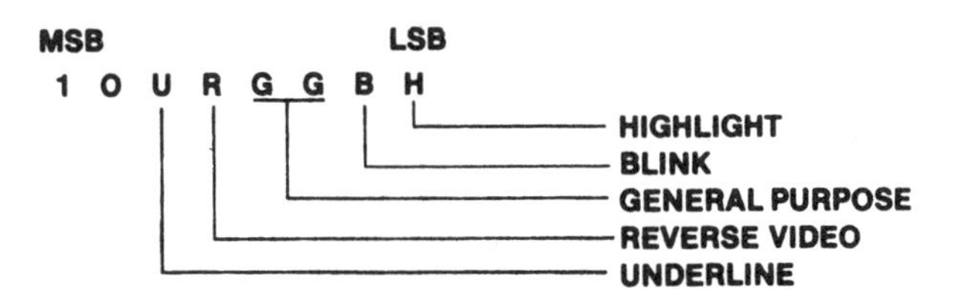

during the vertical retrace interval. The code itself shows up on the screen as a blank character before the first character affected.

The Blink output causes designated characters to blink on and off as the controller pulses the VSP output line. Highlight causes the HGLT output to be activated. Through the external CRT driving circuitry, the character display is intensified.

The Reverse Video output (RVV) is activated by the reverse video attribute. Through the CRT driving circuits, the displayed characters are changed from a light-on-dark presentation to dark-on-light presentation. Underlining occurs when the LTEN (Light Enable) output is asserted. The actual position of the underline, relative to the character, is preprogrammed with the Reset Command. The underline in the character dot matrix depends on the number of scan lines comprising the character.

General purpose outputs (GPA0 & GPA1), controlled by two general purpose attributes, are available for any additional features the designer cares to implement. The outputs can be used to condition CRT driving circuit operation.

Controlling the Cursor Display

The cursor style and the cursor position on the screen can be regulated by two commands. The Reset Command uses the fourth parameter byte to specify a blinking or nonblinking underline or a blinking or nonblinking reverse video block cursor.

The position of the cursor is set by a Cursor Row Register and a Cursor Character Position Register, which are set by the Load Cursor Position Command. The Row Register controls the vertical position of the cursor on the screen while the Character Register controls the horizontal location.

II.13

Bus Arbiter

PART NUMBER	8289/8289-1
FUNCTION	BUS ARBITER
MANUFACTURER	INTEL
VOLTAGES	Vcc = +5
PWR. DISS.	825 mW
PACKAGE	20-pin DIP
TEMPERATURE	0°C → +70°C
FEATURES	FOUR OPERATING MODES COMPATIBLE WITH MULTIBUS BUS STANDARD PROVIDES BIPOLAR BUFFERING AND DRIVE CAPABILITY PROVIDES MULTI-MASTER SYSTEM BUS PROTOCOL
COMPATIBLE MICROPROCESSORS	8086 8088 8089 80186
FUNCTIONAL DESCRIPTION	USED TO INTERFACE MULTIPLE PROCESSORS TO A COMMON BUS STRUCTURE (MULTI-MASTER SYSTEM BUS). THE PROCESSOR OPERATES WITHOUT REGARD TO THE ARBITER'S EXISTENCE IN THE SYSTEM. THEREFORE, CONTROL OF THE BUS IS DETERMINED BY THE ARBITER LOGIC.

Figure II.13.1 Pinout

Courtesy of Intel Corp.

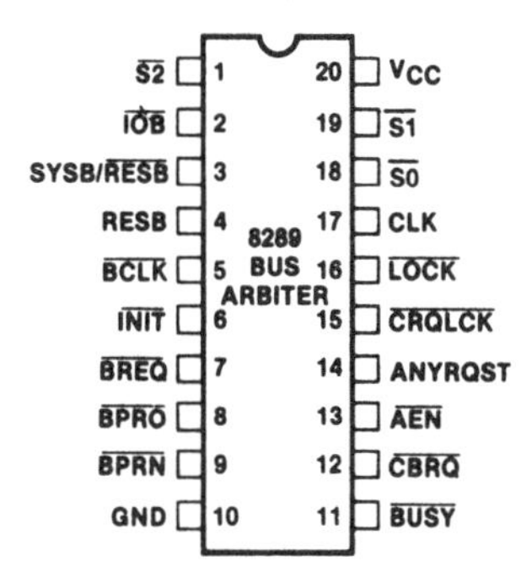

Figure II.13.2 Block Diagram

Courtesy of Intel Corp.

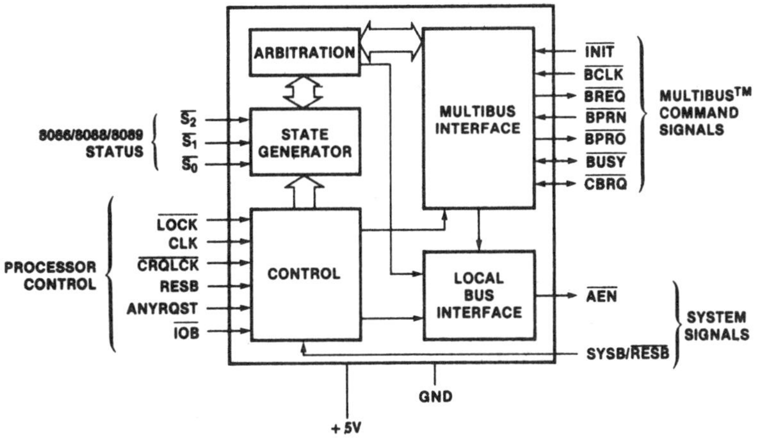

PIN NAME	PIN SYMBOL	PIN NUMBER	FUNCTION
STATUS INPUTS	$\overline{S0}$ $\overline{S1}$ $\overline{S2}$	18 19 1	These three input pins are obtained from the CPU associated with the arbiter. Status lines are internally decoded to provide information on CPU activity.
ADDRESS ENABLE	$\overline{AEN}$	13	Address Enable, when low, is a control signal that forces address latches and Bus Controllers to tri-state their output drivers.
SYSTEM BUS/ RESIDENT BUS	SYSB/$\overline{RESB}$	3	This input function depends on the arbiter operating mode. When activated (see pin 4, RESB), the SYSB/$\overline{RESB}$ pin is selected by address decoding circuits permitting access to a multi-master system bus.
RESIDENT BUS	RESB	4	This pin is physically tied high, allowing the SYSB/$\overline{RESB}$ pin (pin 3) to control the multimaster system bus. When low, pin three is ineffective.
ANY REQUEST	ANYRQST	14	ANYRQST is a strapping option pin used to determine arbiter bus priority.
IO BUS	$\overline{IOB}$	2	Tied low, this pin places the arbiter in a mode which allows it to operate in systems that support an I/O Bus and a multimaster system bus.

PIN NAME	PIN SYMBOL	PIN NUMBER	FUNCTION
COMMON REQUEST	$\overline{\text{CRQLCK}}$	15	This input, when asserted low, prevents the arbiter from giving up the multi-master system bus to another arbiter issuing a bus request via the $\overline{\text{CBRQ}}$ line.
LOCK	$\overline{\text{LOCK}}$	16	$\overline{\text{LOCK}}$ is a CPU-generated signal that prevents the arbiter from surrendering the bus to another arbiter.
CLOCK	CLK	17	This input accepts the system clock driving the CPU and is used to establish the arbiter's internal timing.
INITIALIZE	$\overline{\text{INIT}}$	6	Initialize is a Multibus signal that, when low, resets all bus arbiters. No arbiter will have control of the bus after initialization.
BUS CLOCK	$\overline{\text{BCLK}}$	5	This Multibus clock synchronizes all bus signals.
BUS REQUEST	$\overline{\text{BREQ}}$	7	An arbiter brings this line low requesting access to the multi-master system bus when connected as part of a parallel priority resolving scheme.
BUS PRIORITY IN	$\overline{\text{BPRN}}$	9	$\overline{\text{BPRN}}$ is a low active signal indicating that the arbiter will have access to the bus. Access is obtained on the falling edge of $\overline{\text{BCLK}}$.

(*continued*)

PIN NAME	PIN SYMBOL	PIN NUMBER	FUNCTION
BUS PRIORITY OUT	$\overline{\text{BPRO}}$	8	This signal is used in a serial (daisy-chain) priority resolving scheme.
BUSY	$\overline{\text{BUSY}}$	11	When low, $\overline{\text{BUSY}}$ indicates that the bus is not available. When high, the arbiter with priority may then occupy the bus. This open collector, bidirectional signal is controlled by the arbiter using the bus.
COMMON BUS REQUEST	$\overline{\text{CBRQ}}$	12	$\overline{\text{CBRQ}}$ indicates to an arbiter that lower priority arbiters wish to use the multi-master system bus.
POWER	Vcc	20	+5-V supply.
GROUND	GND	10	Ground.

The basic microprocessor system configuration depicted in Section I of this book shows memory and I/O devices attached to the address and data buses. If a designer wishes to expand this system, he could do so simply by adding more memory or I/O devices to the existing buses (subject to the constraints of the microprocessor chosen). This might lead one to believe that systems of increased complexity and sophistication are all created in this manner, but this is certainly not the case. The system referred to contains a single controlling CPU. Does this also imply that one and only one CPU can control a microprocessor system? The answer is an emphatic no.

Systems with more than one CPU operating with a common bus structure are often referred to as multiprocessing systems. The microprocessors attached in this manner control their own unique environments independently and thereby bring increased processing efficiency to the overall system. These processors most likely will share some common memory or I/O resource and will attempt to access these resources via the common system bus. A bus arbiter synchronizes the processor's access to the common system bus, since only one may control it at any moment in time.

The 8289 (Figs. II.13.1 and II.13.2) is a bus arbiter specifically designed to interface 8086, 8088, 8089, and 80186 microprocessors within a multiprocessor system. Many of the command signals generated by the 8289 are compatible with the Intel MULTIBUS standard.

AN 8289 FUNCTIONAL OVERVIEW

A bus arbiter may or may not operate in association with other external hardware. The VLSI density level of the arbiter chip determines whether or not additional hardware is needed. The 8279 requires some external hardware that will, in turn, control the system buses. Extra hardware is also necessary to support priority resolution.

Each microprocessor in a multiprocessor system has a corresponding arbiter attached to it. The arbiter monitors bus availability, but is essentially transparent to the processor. During the course of normal processing, a microprocessor attempts to access the system buses. Since there are other processors attempting to do this as well, the buses may not be available at this point. The Processor status signals $\overline{S0}$, $\overline{S1}$, and $\overline{S2}$ inform the arbiter of pending processor action and initiate an appropriate arbiter response. If a request for the bus occurs and the bus is not free, then the arbiter will send a Not Ready signal back to the processor (generally through a Clock Generator chip). This forces the processor into wait states which continue until the arbiter is able to acquire control of the bus. At the same time, the arbiter disables all Bus Controllers, transceivers, and address latches that may interfere with ongoing system bus activity.

Eventually the system bus becomes available. The arbiter detects this and releases the processor and bus-related hardware from inactivity, thus allowing a bus transfer to take place.

BUS CONFIGURATIONS SUPPORTED BY THE 8289

The 8289 can, in fact, provide processor access to several bus configurations. The bus primarily used for multiprocessing operations is referred to as a multi-master system bus. The control pins supporting the multi-master system bus conform to the Intel MULTIBUS specification and allow processing devices to have access to the bus. Two other bus types are also supported: an I/O Peripheral Bus and a Resident Bus.

An I/O Peripheral Bus is one in which all attached devices (including memory chips) are treated as I/O devices and are accessed via processor I/O commands. If a memory reference command rather than an I/O command were issued by the processor, the multi-master system bus would automatically be selected for use by the aribter. The I/O bus is primarily used with an I/O processor.

A Resident Bus is dedicated to a single processor or bus master and can handle both memory and I/O commands. The processor on this bus may have access to a multi-master system bus as well. The particular buses described in this section were established by Intel for their product lines but they do reflect the various kinds of buses that occur in complex computer systems.

External bus arbiter strapping pins are tied high or low to select the buses used for a particular arbiter/processor combination. For instance, the $\overline{\text{IOB}}$ pin (I/O Bus), when low, informs the arbiter that the system contains both an I/O Peripheral Bus and a multi-master system bus. The external pin RESB (Resident Bus), when high, informs the arbiter that a Resident Bus and a multi-master system bus are in use. If both of these strapping pins are inactive, then only a multi-master system bus is supported; if both strapping pins are active then all three bus types are functional.

Bus configurations are specified when the system is initially designed. Once specified, the arbiter automatically responds to the processor and routes information to the appropriate bus. The arbiter is made aware of processor requirements by monitoring the processor status lines $\overline{\text{S0}}$, $\overline{\text{S1}}$, and $\overline{\text{S2}}$. Acquiring access to a Resident Bus or an I/O Peripheral Bus is not difficult since these buses are dedicated to a processor. When the same processor requires access to the multi-master system bus, it may have to wait until the arbiter can establish a connection to it. The arbiter's ability to do this depends on the current availability of the multi-master system bus, the arbiter's priority relative to the priority of other arbiters in the system, and other special conditions such as CPU Lock.

Priority Resolution Techniques

The means by which a processor obtains the multi-master system bus depends upon the priority resolution technique in use. Priority can be assigned to the 8279 arbiters in one of three possible ways. Serial priority is the easiest to implement because external logic is not required. The arbiters are placed in a daisy-chain configuration by connecting the Bus Priority Out ($\overline{\text{BPRO}}$) line of one chip to the Bus Priority In ($\overline{\text{BPRN}}$) line of the next chip in the chain. The arbiter chip position in the chain determines relative priority. (The chip with the highest priority has its $\overline{\text{BPRN}}$ line tied to ground.) Any device requiring service brings its $\overline{\text{BPRN}}$ line high, thus informing all lower priority devices that a higher priority request has been made. Since the action of any arbiter depends upon the number of arbiters in the chain, the propagation delay time involved constrains the usefulness of this technique. The signal must wind its way through the daisy-chain in a period of time less than that of the multi-master system clock ($\overline{\text{BCLK}}$). For example, with a 10 MHz $\overline{\text{BCLK}}$, only three arbiters may be daisy-chained together.

Parallel Priority resolution can support more arbiters than can the serial technique, but it requires external logic circuits. These circuits consist of priority encoder and decoder circuits attached to the arbiters through the Bus Request ($\overline{\text{BREQ}}$) and Bus Priority Input ($\overline{\text{BPRN}}$) lines. An arbiter desiring bus access activates its $\overline{\text{BREQ}}$ line. The arbiter's priority depends upon the way it is physically attached to the priority resolution hardware. If the arbiter requesting access is the arbiter with the greatest priority at the time (other, lower priority arbiters could also request the bus at the same time), then its $\overline{\text{BPRN}}$ line will be asserted. This indicates that the bus will become available to the arbiter when current bus activity ceases.

Rotating Priority is similar to Parallel Priority. In this case, priority rotates among the various arbiters so that over a period of time all have equal access to the bus. The external circuitry required is more complex than that required for Parallel Priority.

Two other signal lines are necessary for a complete priority system. Both Common Bus Request ($\overline{\text{CBRQ}}$) and $\overline{\text{BUSY}}$ are open-collector lines. The BUSY pins from each arbiter are wire-ORed together. The same is true for the $\overline{\text{CBRQ}}$ pins (Figs. II.13.3 and II.13.4). CBRQ is pulled low when an arbiter desires access to the bus. All other arbiters, including the one currently accessing the bus, can sense this line. If all proper conditions exist, the processor/arbiter combination currently on the bus relinquishes control when the present transfer cycle is complete. Furthermore, a processor/arbiter using the bus may maintain possession of the bus if $\overline{\text{CBRQ}}$ is inactive. This indicates that no other device wishes to use the bus. Processor/arbiter overhead involved in accessing the bus is eliminated when this condition is detected.

The $\overline{\text{BUSY}}$ line is pulled low by the arbiter with priority when it finally gains access to the multi-master system bus. This occurs as soon

Figure II.13.3 Parallel Priority Resolving Technique
Courtesy of Intel Corp.

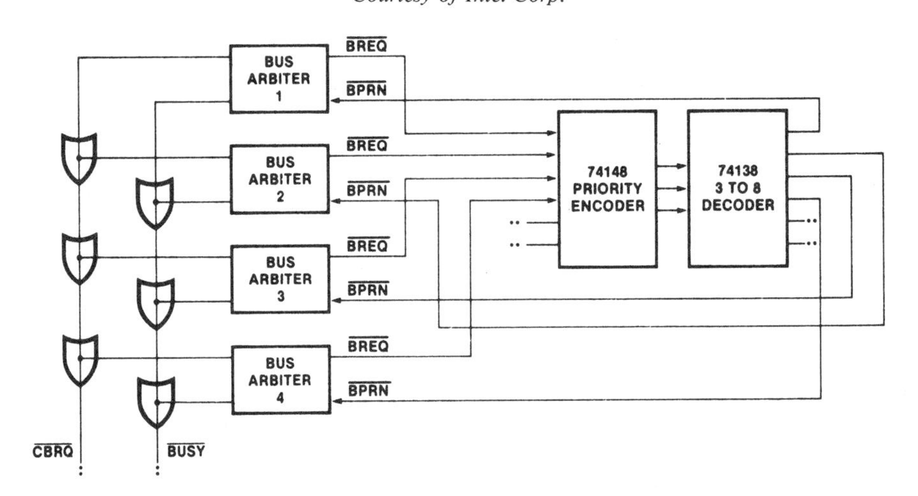

as the current user of the bus relinquishes control by deactivating $\overline{\text{BUSY}}$. All other arbiters sense this condition and are prevented from using the bus.

It should be noted that the various bus control signals are synchronized to the bus clock, $\overline{\text{BCLK}}$.

Figure II.13.4 Serial Priority Resolving Technique
Courtesy of Intel Corp.

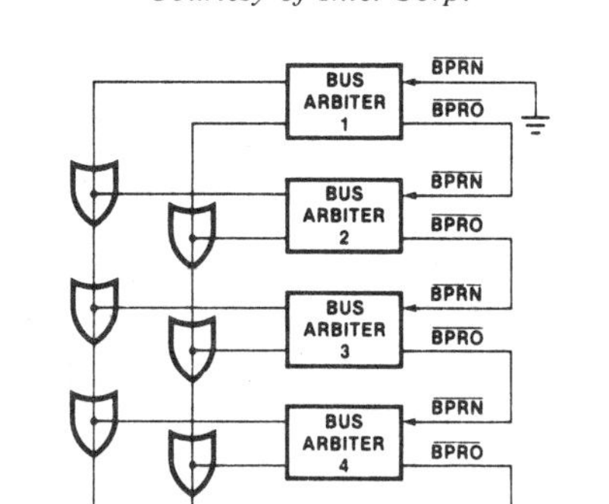

THE NUMBER OF ARBITERS THAT MAY BE DAISY-CHAINED TOGETHER IN THE SERIAL PRIORITY RESOLVING SCHEME IS A FUNCTION OF BCLK AND THE PROPAGATION DELAY FROM ARBITER TO ARBITER. NORMALLY, AT 10 MHz ONLY 3 ARBITERS MAY BE DAISY-CHAINED.

Understanding the Multi-Master System Bus Control Sequence

Acquiring and giving up access to the multi-master system bus depends upon a number of conditions. Foremost is the mode of operation. If the arbiter is configured for multi-master operation only, then the bus will be requested whenever a valid bus request is made. This includes both memory and I/O operations. If, on the other hand, a Resident Bus (RESB = 1) and/or an I/O Peripheral Bus ($\overline{\text{IOB}}$ = 0) is also a part of the bus arbiter's configuration, then control line information as well as the processor bus operation in progress determines which bus is used. For the case of an I/O Peripheral Bus, only memory operations will be routed to the multi-master system bus. Any I/O operations automatically use the I/O Bus. A control line called SYSB/$\overline{\text{RESB}}$ (System Bus/Resident Bus), when high, selects the multi-master system bus for use. SYSB/$\overline{\text{RESB}}$ is itself controlled by memory-mapping circuitry that determines which bus to access.

In addition to the bus designation logic mentioned above, the sequence of events that must occur for a bus request to actually take place are:

1. The arbiter senses activity within its associated processor on the status lines $\overline{\text{S0}}$, $\overline{\text{S1}}$, and $\overline{\text{S2}}$.
2. The arbiter establishes its priority by raising its $\overline{\text{BPRN}}$ line (serial priority) or by lowering $\overline{\text{BREQ}}$ (parallel priority). In addition, the $\overline{\text{CBRQ}}$ (Common Bus Request) line is also asserted.
3. If the arbiter has the highest priority ($\overline{\text{BPRN}}$ line is active), when the $\overline{\text{BUSY}}$ line goes high the arbiter takes over the bus by bringing $\overline{\text{BUSY}}$ back low. Other arbiters are then prevented from using the bus.
4. $\overline{\text{AEN}}$ (Address Enable) is activated to allow the Bus Controllers and address latches to electrically access the multi-master system bus.

Surrendering the bus depends on the mode of operation, the priority, and the processor status. A higher priority request normally forces a lower priority processor off the bus as soon as the lower priority bus completes its operation. If the processor currently on the bus enters the HALT or Idle states, the bus is also surrendered.

The arbiter can maintain control of the bus, even if other higher priority requests are made, through the use of the $\overline{\text{LOCK}}$ or Common

Request Lock ($\overline{\text{CRQLCK}}$) inputs. $\overline{\text{LOCK}}$ is a processor output that, when active, prevents the arbiter from surrendering the bus. Common Request Lock, when active, prevents the arbiter from surrendering the bus to other arbiters making a bus request via $\overline{\text{CBRQ}}$. These lines should be used judiciously to prevent tying up the bus for extended periods of time. Reserving the bus with these signals is typically done for critical, real-time processes. An example is the transfer of a multibyte interrupt vector across the bus. It would be imprudent to give up access to the bus midway during the transfer since the interrupt would be held up. The processor undergoing the interrupt can generate $\overline{\text{LOCK}}$ to maintain bus access until the interrupt vector transfer is completed.

Finally, the priority system may also be altered with the ANYRQST strapping option. This pin, if strapped high, forces higher priority arbiters to surrender the bus to lower priority arbiters upon an active $\overline{\text{CBRQ}}$. The bus is relinquished as soon as the current process is complete. If $\overline{\text{CBRQ}}$ is tied low and ANYRQST is tied high, then the bus will be automatically surrendered at the end of each transfer cycle.

II.14

Data Encryption Unit

PART NUMBER	TMS7500 TMS75C00
FUNCTION	DATA ENCRYPTION DEVICE
MANUFACTURER	TEXAS INSTRUMENTS
VOLTAGES	Vcc = +5 NOMINAL
PWR. DISS.	400 mW
PACKAGE	40-pin DIP
TEMPERATURE	0°C → +70°C
FEATURES	PERFORMS NATIONAL BUREAU OF STANDARDS DATA ENCRYPTION STANDARD (DES) CMOS (TMS75C00) VERSION AVAILABLE ELECTRONIC CODEBOOK AND CIPHER FEEDBACK MODES 64-BIT MASTER AND ACTIVE KEY STORAGE
COMPATIBLE MICROPROCESSORS	8-BIT PROCESSORS (EXTERNAL HARDWARE REQUIRED)
FUNCTIONAL DESCRIPTION	THE DATA ENCRYPTION DEVICE CONVERTS DATA INTO A CODED FORM THAT IS UNREADABLE TO UNAUTHORIZED USERS. THE DATA IS TRANSFORMED INTO CIPHER TEXT UTILIZING A USER-DEFINED KEY AND THE DATA ENCRYPTION STANDARD (DES) ALGORITHM. THE RESULTING CIPHER TEXT IS EXTREMELY DIFFICULT TO DECODE WITHOUT PRIOR KNOWLEDGE OF THE ORIGINAL USER KEY.

Figure II.14.1 TMS7500 Pinout

Courtesy of Texas Instruments, Inc.

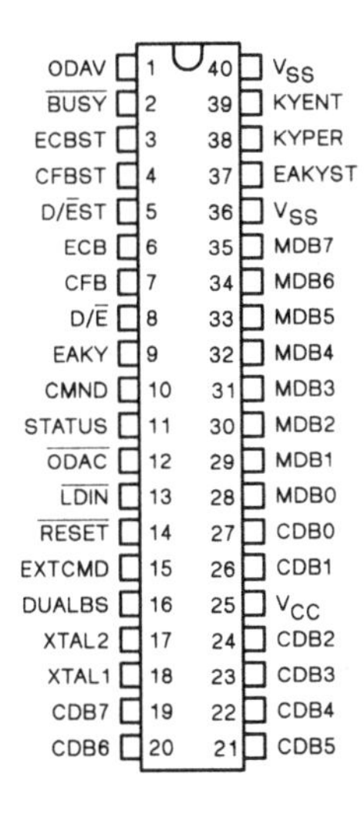

Figure II.14.2 TMS7500 Block Diagram

Courtesy of Texas Instruments, Inc.

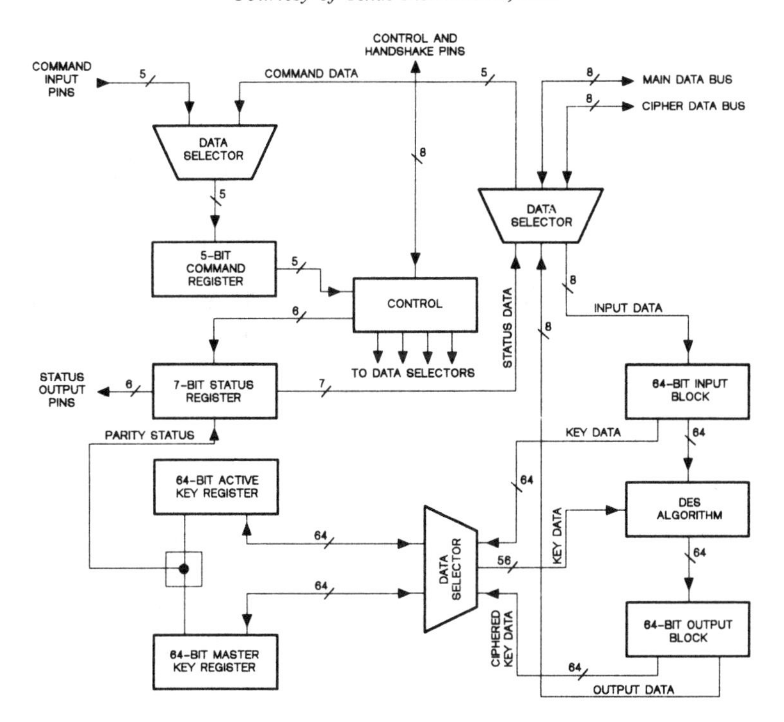

PIN NAME	PIN SYMBOL	PIN NUMBER	FUNCTION
CRYSTAL INPUT 1 CRYSTAL INPUT 2	XTAL1 XTAL2	18 17	A crystal or ceramic resonator (5.0 MHz maximum) sets the Data Encryption Device (DED) clock frequency.
OUTPUT DATA AVAILABLE	ODAV	1	The DED informs external controlling hardware that data is available by raising this control line high.
OUTPUT DATA ACCEPTED	$\overline{\text{ODAC}}$	12	This line is brought low by external controlling hardware indicating that data has been read from the DED.
LOAD DATA IN	$\overline{\text{LDIN}}$	13	$\overline{\text{LDIN}}$ is asserted low by the controlling CPU hardware to initiate a write operation. Data may be written into the DED on either the Main Data Bus or the Cipher Bus.
BUSY	$\overline{\text{BUSY}}$	2	$\overline{\text{Busy}}$ is pulled low by the DED in response to LDIN. When $\overline{\text{Busy}}$ is active, the DED is storing data presented to it by the host CPU.
DUAL DATA BUS	DUALBS	16	This pin selects when the Cipher Data Bus will be used. When high, both the Main Data Bus and the Cipher Data Bus may be used. When low, data transmission can take place

PIN NAME	PIN SYMBOL	PIN NUMBER	FUNCTION
			only on the Main Data Bus.
CIPHER DATA BUS 0	CDB0	27	This data bus is used for the transmission of encrypted data. It is placed into operation when DUALBS is high.
CIPHER DATA BUS 1	CDB1	26	
CIPHER DATA BUS 2	CDB2	24	
CIPHER DATA BUS 3	CDB3	23	
CIPHER DATA BUS 4	CDB4	22	
CIPHER DATA BUS 5	CDB5	21	
CIPHER DATA BUS 6	CDB6	20	
CIPHER DATA BUS 7	CDB7	19	
MAIN DATA BUS 0	MDB0	28	The Main Data Bus is used for all data exchanges when DUALBS is low. If DUALBS is high, then both the Cipher Bus and Main Data Bus are functional.
MAIN DATA BUS 1	MDB1	29	
MAIN DATA BUS 2	MDB2	30	
MAIN DATA BUS 3	MDB3	31	
MAIN DATA BUS 4	MDB4	32	
MAIN DATA BUS 5	MDB5	33	
MAIN DATA BUS 6	MDB6	34	
MAIN DATA BUS 7	MDB7	35	
COMMAND REGISTER UPDATE	CMND	10	When high, CMND enables Command Register data input. Input EXTCMD determines the command information source.
EXTERNAL COMMAND	EXTCMD	15	Command data is present at chip input pins when

(continued)

PIN NAME	PIN SYMBOL	PIN NUMBER	FUNCTION
			EXTCMD is high. When low, command data is extracted from the Main Data Bus.
ELECTRONIC CODEBOOK	ECB	6	This input pin sets the Command Register for ECB mode of operation when EXTCMD and CMND are high.
CIPHER FEEDBACK	CFB	7	This input pin sets the Command Register for CFB mode of operation when EXTCMD and CMND are high.
DECRYPT/ENCRYPT	$D/\overline{E}$	8	This input pin sets a corresponding bit in the Command Register. $D/\overline{E}$ determines data flow direction along the data buses and controls decrypt/encrypt operation.
ENTER ACTIVE KEY	EAKY	9	The EAKY bit in the Command Register may be set using this input pin.
READ STATUS	STATUS	11	Activating this input (high) enables the DED to place status information onto the Main Data Bus. Status information is also available on the following output pins:
ELECTRONIC CODEBOOK STATUS	ECBST	3	Status output pin.

PIN NAME	PIN SYMBOL	PIN NUMBER	FUNCTION
CIPHER FEEDBACK STATUS	CFBST	4	Status output pin.
DECRYPT/ENCRYPT STATUS	D/$\overline{\text{E}}$ST	5	Status output pin.
ENTER ACTIVE KEY STATUS	EAKYST	37	Status output pin.
KEY PARITY ERROR	KYPER	38	Status output pin.
KEY ENTERED	KYENT	39	Status output pin.
RESET	$\overline{\text{RESET}}$	14	This active low line clears the Status, Command, and Key Registers. Both data buses are placed into a high impedance state after reset.
POWER SOURCE	Vcc	25	4.5 V to 5.5 V for the TMS7500 and 3 V to 5.5 V for the TMS75C00.
POWER GROUND	Vss	36	Ground.
POWER GROUND	Vss	40	Ground.

Data encryption is a process by which usable data, called plain text, is transformed into a coded form, called cipher text. Cipher text is essentially unusable to those unauthorized to access it. Encrypted data, or ciphered data, can be produced in many ways. The method used in the TMS7500 Data Encryption Device (DED) chip conforms to the National Bureau of Standards Data Encryption Standard (DES) algorithm. This standard is set forth in Federal Information Processing Standard Publication 46.

The DES algorithm requires the use of a 64-bit, user-specified key (56 bits plus parity). Data security depends on the confidentiality of this key. Sixty-four-bit plain text data is combined with the user key, using the DES standard, to produce 64 bits of cipher text data. The cipher text bears no resemblance to the original data and can be transmitted with a high degree of security. A receiving DED chip combines the cipher text with the same user key, reconstituting the original plain text data. The same DES algorithm is used in this case as well. The DES algorithm is a matter of public knowledge and cannot, by itself, insure the secrecy of the data it encrypts. The secrecy of the user key must be maintained for data protection. Current estimates suggest that to illegally decode cipher text, by trying all possible combinations of the lengthy user key, would take several years on a modern computing system.

TMS7500 FUNCTIONAL OVERVIEW

The TMS7500 and TMS75C00 are NMOS and CMOS versions of the same chip. Some differences in voltage levels, power dissipation, and timing set the two apart. Both will be referred to as TMS7500 or DED in the discussion to follow. Figures II.14.1 and II.14.2 show the pinout and functional block diagram of the TMS7500.

Designed into the DED are several registers used to store commands, status, and keys. All internal DED activities are controlled by the Command Register (Figure II.14.3). Placing the individual bits to desired levels (a high is active for all bits) initiates a DED action. The Command Register may be loaded either from external chip pins that match Command Register bit function or through a data transfer on the Main Data Bus. The Status Register is similar to the Command Register because its contents are also accessible from the Main Data Bus or external chip pins. The flexibility gained from these two options allows for either software or hardware control of the DED chip.

A Master Key Register and an Active Key Register provide simultaneous storage of two 64-bit keys. The Active Key Register is used for normal data encryption. The Master Key Register can be used to encrypt or decrypt the Active Register Key.

Two 8-bit data buses can be used to form a CPU interface. The Main Data Bus is used for the majority of information exchange be-

Figure II.14.3 Command Register

Courtesy of Texas Instruments, Inc.

RESET2	X	X	X	EAKY	D/$\overline{E}$	CFB	ECB
MSB	(X = DON'T CARE)						LSB

tween the CPU and DED. Commands, Status, Plain Text, and Cipher Text may all be found on the Main Data Bus. The optional Cipher Data Bus is used primarily for Ciphered Text and, since it is separate from the common System (Main) Data Bus, offers additional security for data transferrals. Both the Main and Cipher Data Buses are controlled by a common set of handshaking signals and, in some modes, can function at the same time.

UNDERSTANDING TMS7500 OPERATION

A reset, provided by holding the $\overline{\text{RESET}}$ input low, clears all registers and the Status output pins. A reset should be performed when the chip is first powered up. The Master Key Register should be the first register loaded following a reset. Since the Data Buses are only eight bits wide, eight write operations are necessary to completely load the 64-bit key register (either bus can be used). The Active Key Register is loaded along with the Master Key Register. The Active Key Register can be loaded at any time, but, until the Master Key is loaded, no encryption/decryption operations can begin.

Data writes to the Master Key Register (or any register) are guided by two handshaking signals, Load Data In ($\overline{\text{LDIN}}$) and $\overline{\text{BUSY}}$. Lowering the $\overline{\text{LDIN}}$ line signals the chip that an information byte is present on one of the data buses. (The control signals that differentiate between bus and register use are discussed later.) The DED responds by lowering $\overline{\text{BUSY}}$, indicating that the information byte has been accepted and stored. When $\overline{\text{BUSY}}$ returns to a high level, the DED is ready for another write or read cycle. It should be noted that a write operation is ignored by the DED when output data is available.

Once the Master Key is loaded, a Command Register write operation initiates one of the DED functions. A Command Register write is distinguished from others by the fact that the Command Register Update (CMND) input is high. When this line is active, the data sent to the DED is automatically set into the Command Register. Two options are possible here: the information can come directly from the Main Data Bus or from the input pins corresponding to Command Register bits. These individual bits control DED operation and are discussed later. Control line External Command (EXTCMD) determines the source of command data. When high, command information is obtained from the input pins provided, whereas a low indicates that the Main Data Bus supplies the information.

As DED processing proceeds, data or chip status may be read.

Two additional handshaking signals are designated for this purpose. Output Data Available (ODAV) goes high when data or status is ready to be read. The host CPU then informs the DED that this information has been accepted by sending a negative pulse on the Output Data Accepted ($\overline{\text{ODAC}}$) line. Status information can be read, provided that a request for status information was made prior to ODAV becoming active. The request is made by bringing the Status input pin high. This request can be made at any time, but may be held up if other bus activity is in progress. Once the status is made available, all other chip activity is suspended, so a request for status should immediately be followed by the proper read operation. The Status Register can be read out on the Main Data Bus. Additionally, all Status Register information, except for the Message Start (MSGST) bit, is also available on dedicated output pins. These pins may be read at any time but require auxiliary hardware to integrate them into overall computer system operation.

Using the TMS7500 Command and Status Registers

A software reset is initiated by setting the most significant Command Register bit (Fig. II.14.3). A software reset has the same effect as a hardware reset. That is, all register and external Status pins are cleared.

The two least significant Command Register bits initiate different data encryption and decryption methods. Only one of the two may be active at a time. Cipher Feedback (CFB) and Electronic Codebook (ECB) are detailed in the data encryption section.

The Enter Active Key (EAKY) bit is set prior to loading the Active Key Register. Once set, the next eight write operations load the Active Key Register. After these write cycles are complete, the EAKY bit must be cleared to prevent any further writes from altering the Active Key Register.

The final bit, Decrypt/Encrypt (D/$\overline{\text{E}}$), has several functions. First, it determines if a data encryption (D/$\overline{\text{E}}$ = 0) or decryption (D/$\overline{\text{E}}$ = 1) process will take place. Second, the bit indicates the direction of data flow when both the Main Data Bus and the Cipher Bus are in use.

Input pin Dual Data Bus (DUALBS), when high, allows both data buses to function. Figure II.14.4 indicates the read/write capabilities of the two buses as well as the D/$\overline{\text{E}}$ Command Register bit use in determining data transfer direction. Recall that when reading or writing, the handshaking signals already discussed affect both data buses.

Figure II.14.4 TMS7500 Data Flow
Courtesy of Texas Instruments, Inc.

DUALBS PIN	D/$\overline{E}$ BIT	CIPHER DATA BUS	MAIN DATA BUS
0	0	NOT USED	READ/WRITE
0	1	NOT USED	READ/WRITE
1	0	READ FROM	WRITE TO
1	1	WRITE TO	READ FROM

The Status Register and the status output pins indicate the condition of DED operations. Figure II.14.5 details the individual bits within the Status Register. The four least significant bits, Enter Active Key Status (EAKYST), Decrypt/Encrypt Status (D/$\overline{E}$ST), Cipher Feedback Status (CFBST), and Electronic Codebook Status (ECBST), simply reflect the state of their corresponding bits in the Command Register.

The bit labeled KYENT (Key Entered) is set after the eight bytes of Master Key Data have been stored. Recall that further chip operation is not possible until the Master Key is in place, so this status bit provides a handy indication of this critical activity.

Key Parity Error (KYPER) is set if an odd parity error occurs during a key load. The key is actually 56 bits long with each byte containing seven bits of key information and one parity bit. Parity is checked after all eight bytes have been loaded. This status bit is cleared upon reset or when a new key with correct parity is loaded.

The final bit, Message Start (MSGST), is the only status bit without a corresponding status output pin. This bit is set when the Command Register bit for Cipher Feedback is also set. The MSGST bit indicates that the next eight data bytes written to the DED are designated as the Initialization Vector (see modes section that follows) for the Cipher Feedback process. MSGST is cleared when all eight bytes have been written, when a reset occurs, or when another operational mode is selected.

Figure II.14.5 Status Register
Courtesy of Texas Instruments, Inc.

Status Register

0	KYENT	MSGST	KYPER	EAKYST	D/$\overline{E}$ST	CFBST	ECBST
MSB							LSB

Data Encryption/Decryption Modes of Operation

Two data protection methods are possible with the DED chip. The Electronic Codebook (ECB) method takes 64 bits of plain text input and produces 64 bits of ciphered text output.

This mode is initiated by setting the Command Register ECB bit. Each eight-byte block of data written to the chip is acted upon by the DED algorithm. The input data can be encrypted or decrypted as stipulated by the $D/\overline{E}$ Command Register bit. The resulting output also depends upon the key stored in the Active Key Register. All data transmission takes place on the Main Data Bus if only the Main Data Bus is selected for use. When both buses are enabled (DUALBS = 1), then plain text flows along the Main Data Bus and cipher text is transmitted along the Cipher Bus.

Another method allows one data byte to be encrypted/decrypted at a time. This method, Cipher Feedback, is useful for low speed, serial data applications. Cipher Feedback requires an Initialization Vector (IV) which is loaded as described in the Status register section. The IV is 64 bits long and is the first item written to the chip when the Cipher Feedback mode is selected. Cipher Feedback makes use of a pseudorandom string of binary numbers generated by the DED. This stream of numbers is Exclusive-ORed with plain text data creating the cipher text. (Initially, the IV, rather than plain text is used.) The cipher text is then fed back and Exclusive-ORed with a section of plain text. Once past the initialization stage, each input byte of data is encrypted or decrypted in this fashion and the results are output. Bus usage is the same as that in Electronic Codebook mode.

The organization of 64-bit data presented to the chip and received from the chip is as follows: The most significant bit of the first byte becomes bit one of the 64-bit string; the least significant bit of the eighth byte becomes bit 64 of the 64-bit string.

When neither Electronic Codebook nor Cipher Feedback modes are initiated, it becomes possible for plain text to move unaltered through the chip. If both data buses are enabled, plain text may be passed from one bus to the other under control of the $D/\overline{E}$ Command Register bit.

II.15

Error Detection and Correction Unit

PART NUMBER	2960 Am2960/Am2960A
FUNCTION	ERROR DETECTION AND CORRECTION (EDC) UNIT
MANUFACTURER	SIGNETICS ADVANCED MICRO DEVICES
VOLTAGES	Vcc = +5
PWR. DISS.	1.89 W
PACKAGE	48-pin DIP
TEMPERATURE	0°C → +70°C
FEATURES	EDC CAN INCREASE MEMORY SYSTEM RELIABILITY 60-FOLD CORRECTS ALL SINGLE-BIT ERRORS DETECTS ALL DOUBLE-BIT ERRORS AND SOME TRIPLE-BIT ERRORS CASCADABLE TO HANDLE 8-, 16-, 32-, OR 64-BIT EDC BUILT-IN DIAGNOSTICS
COMPATIBLE MICROPROCESSORS	MOST PROCESSORS - SOME EXTERNAL CIRCUITS REQUIRED
FUNCTIONAL DESCRIPTION	EDC CIRCUITRY, PLACED IN THE DATA PATH BETWEEN MEMORY AND A MICROPROCESSOR, CAN PROVIDE IMPROVED MEMORY RELIABILITY. INTERNALLY, THE IMPLEMENTATION OF A MODIFIED HAMMING CODE RESULTS IN THE CORRECTION OF ALL SINGLE-BIT ERRORS INDUCED INTO PREVIOUSLY STORED DATA. IN ADDITION, THE PRESENCE OF ALL DOUBLE-BIT AND SOME TRIPLE-BIT ERRORS WILL BE DETECTED. CHECK BITS MUST BE STORED ALONG WITH THE ORIGINAL DATA FOR ERROR DETECTION AND CORRECTION TO FUNCTION.

Figure II.15.1 2960 Pinout

Courtesy of Signetics

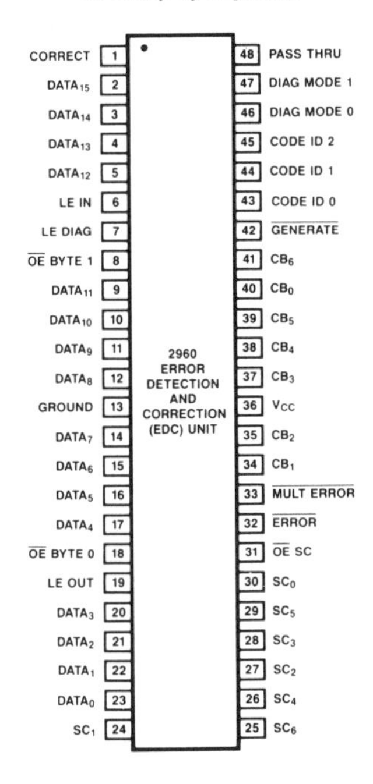

Figure II.15.2 2960 Block Diagram

Courtesy of Signetics

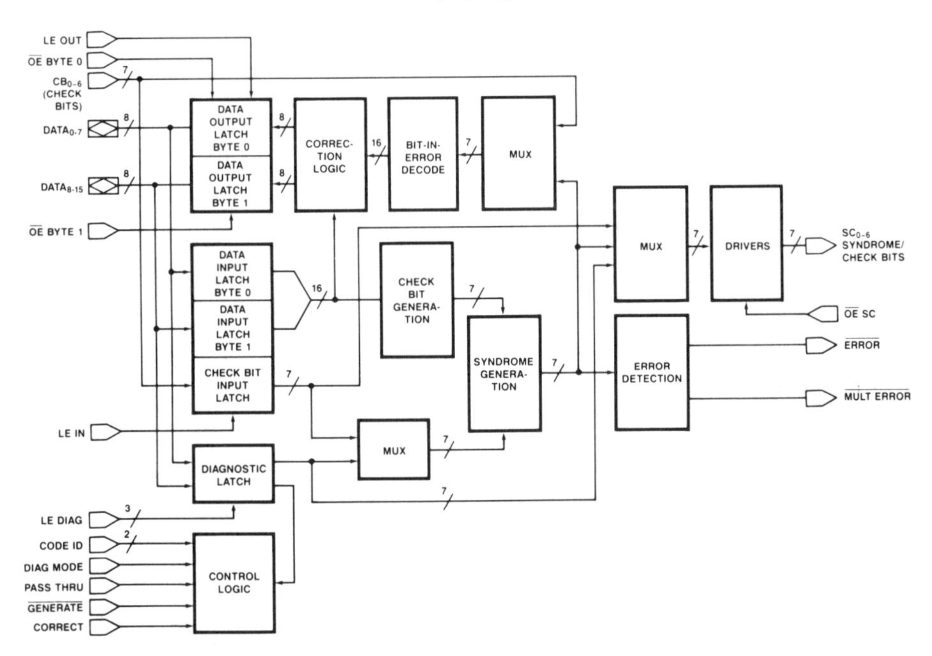

PIN NAME	PIN SYMBOL	PIN NUMBER	FUNCTION
LATCH ENABLE DATA INPUT	LE IN	6	This control line, when high, allows both the Data Input Latch and the Check Bit Input Latch to follow the corresponding data and check bit inputs. The latches hold the input information when LE IN is brought low.
LATCH ENABLE DIAGNOSTIC LATCH	LE DIAG	7	When high, this control line allows the Diagnostic Latches to follow the information on the data line inputs. A low level will latch data into the Diagnostic Latches.
LATCH ENABLE DATA OUTPUT LATCH	LE OUT	19	A low level on LE OUT latches data into the Data Output Latch. A high level causes the latch output to follow its input.
OUTPUT ENABLE 0	$\overline{OE}$ BYTE 0	18	Low levels on these two control lines enable the Data Output Latches. A high level forces the outputs into a high impedance state.
OUTPUT ENABLE 1	$\overline{OE}$ BYTE 1	8	
OUTPUT ENABLE SYNDROME/CHECK BITS	$\overline{OE}$ SC	31	The Syndrome/Check bit outputs are enabled by a low on this control line. A high places the outputs into a high impedance state.
SYNDROME/CHECK BIT 0	SC0	30	Lines SC0-SC6 hold Syndrome bits in Detect or

PIN NAME	PIN SYMBOL	PIN NUMBER	FUNCTION
SYNDROME/CHECK BIT 1	SC1	24	Correct Modes of operation, or Check bits when in Generate Mode.
SYNDROME/CHECK BIT 2	SC2	27	
SYNDROME/CHECK BIT 3	SC3	28	
SYNDROME/CHECK BIT 4	SC4	26	
SYNDROME/CHECK BIT 5	SC5	29	
SYNDROME/CHECK BIT 6	SC6	25	
CHECK BIT IN 0	CB0	40	These lines receive the Check bits as input for error detection functions or Syndrome bits as input when using 32- or 64-bit data.
CHECK BIT IN 1	CB1	34	
CHECK BIT IN 2	CB2	35	
CHECK BIT IN 3	CB3	37	
CHECK BIT IN 4	CB4	38	
CHECK BIT IN 5	CB5	39	
CHECK BIT IN 6	CB6	41	
DATA LINE 0	DATA0	23	These sixteen bidirectional data lines present information to the EDC as well as receive information from the EDC Output Latches.
DATA LINE 1	DATA1	22	
DATA LINE 2	DATA2	21	
DATA LINE 3	DATA3	20	
DATA LINE 4	DATA4	17	
DATA LINE 5	DATA5	16	
DATA LINE 6	DATA6	15	
DATA LINE 7	DATA7	14	
DATA LINE 8	DATA8	12	
DATA LINE 9	DATA9	11	
DATA LINE 10	DATA10	10	
DATA LINE 11	DATA11	9	
DATA LINE 12	DATA12	5	
DATA LINE 13	DATA13	4	
DATA LINE 14	DATA14	3	
DATA LINE 15	DATA15	2	

(*continued*)

PIN NAME	PIN SYMBOL	PIN NUMBER	FUNCTION
CORRECT	COR-RECT	1	Single-bit errors are corrected when this line is high. When the line is low, data errors are detected but not corrected.
PASS THRU INPUT	PASS THRU	48	Both Check bits and data pass through the EDC without modification when this input is high.
GENERATE CHECK BIT	$\overline{\text{GENER-ATE}}$	42	A low level on this line places the EDC into Generate Mode. In this mode, Check bits are generated to reflect the data contents in the Data Input Latch. A high level places the EDC into Detect or Correct Mode.
CODE IDENTIFICATION 0	CODE ID0	43	The three Code ID inputs are set to a specific encoded number to inform the EDC chip or chips of the data size, Hamming Code, or special operations under which they will function.
CODE IDENTIFICATION 1	CODE ID1	44	
CODE IDENTIFICATION 2	CODE ID2	45	
DIAGNOSTIC MODE SELECT 0	DIAG MODE0	46	The diagnostic capability of the chip is controlled by these two inputs.
DIAGNOSTIC MODE SELECT 1	DIAG MODE1	47	
ERROR DETECTED	$\overline{\text{ERROR}}$	32	When the EDC is in the Detect or Correct Mode of operation, this output goes low if errors are discovered. The line is forced

PIN NAME	PIN SYMBOL	PIN NUMBER	FUNCTION
			high in Generate Mode. Additional circuitry is required to use this line with 64-bit data.
MULTIPLE ERRORS	$\overline{\text{MULT ERROR}}$	33	When the EDC is in the Detect or Correct Mode of operation, this output goes low if two or more errors are discovered. The line is forced high in Generate Mode. Additional circuitry is required to use this line with 64 bit data.
+5-V POWER SUPPLY	Vcc	36	+5-V DC input.
GROUND	GND	13	Ground.

As modern day memory system designs increase in size and complexity, the likelihood of data errors also increases. Errors can occur upon retrieval, as data is read from a memory chip. Noise glitches and other disturbances can alter the state of one or more data bits, resulting in eventual processing errors. Internal to the memory chips, both hard and soft errors are possible. A hard error is a failure in one of the memory array storage cells. It is likely that the cell will end up in a permanent stuck-at-one or stuck-at-zero condition, making the cell, and consequently the storage address in which the cell resides, potentially useless. Soft errors occur when a cell changes state temporarily. An alpha particle, for instance, could impart enough energy to a cell to cause a state change. The effect is not permanent, but will result in temporary data errors. Often a multitude of soft errors precede an eventual hard error.

All of the above phenomena result in decreased memory system reliability and contribute to processing errors. Obviously errors must be detected to prevent computations with invalid data. For this reason, many detection techniques are used such as parity checking, checksums, and cyclic redundancy checks. These methods prevent errors from affecting system processing but do not improve system reliability. If, for instance, a hard error crops up in a memory chip, the chip is, for all practical purposes, no longer functional.

The memory system reliability could be improved if errors such as this are automatically corrected as well as detected. The Error Detection and Correction (EDC) chip can do this. Figure II.15.1 is the pinout and Figure II.15.2 the block diagram for the 2960 Error Detection and Correction Unit.

HOW ERRORS ARE DETECTED AND CORRECTED IN EDC LOGIC

Error correction hardware is designed according to mathematical algorithms called Hamming codes. By implementing these codes in hardware, simultaneous data errors can be detected and possibly corrected. The number of errors that can be simultaneously detected/corrected depends upon the actual Hamming code implementation. To detect and correct an increased number of errors, a corresponding increase in EDC hardware is required, so eventually the hardware requirements overwhelm the detecting and correcting power gained. The 2960 has the following capabilities: Any single-bit error, that is, an error in one bit of a data word or quantity of data, can be corrected. Additionally, the existence of all double-bit errors and some triple-bit errors is identified.

Besides the EDC hardware required, a certain amount of additional memory storage is necessary in an error correction/detection system. This additional memory holds the Check bits that are stored with the data. A quantity of data destined for storage has several Check bits generated by the EDC. For example, a 16-bit quantity of data will necessitate six Check bits in an EDC scheme. When any 16-bit data word is stored by a microprocessor instruction, 22 bits will actually go to the memory (16 bits of data and 6 Check bits). Larger quantities of data require additional Check bits. A single 2960 will handle 16 bits of data. Two chips produce seven Check bits for a 32-bit data quantity (4 bytes) while four EDCs produce eight Check bits for a 64-bit quantity. (Some additional hardware is needed for the 8-byte system.) It may appear that the number of Check bits required is substantial, especially for smaller quantities of data. Although this is true, the manufacturer claims that a 60-fold increase in dynamic RAM memory system reliability is possible. Therefore, for many applications, the Check bit overhead is worthwhile.

The Check bits are really nothing more than multiple parity bits generated from the same data. For the 16-bit case, six parity checks are made on different portions of the same data quantity. The data bits used to generate the parity bits are derived from the Hamming codes. The resulting parity bits are then referred to as Check bits and will accompany the data to memory storage during a write cycle. Figure II.15.3 illustrates this segment of the data checking process. The Check bits will reside in the same memory address as the data.

When the CPU issues a data read instruction, both data and Check bits are retrieved from memory and presented to the EDC chip (Fig. II.15.4). The EDC chip reprocesses the data in exactly the same fashion as it did when the original Check bits were generated. If the data is unchanged, then the new Check bits generated match the originals. The two sets of Check bits are compared to verify the data integrity. If no errors are discovered, then the data read operation continues.

The two sets of Check bits are Exclusive-ORed on a bit-by-bit basis. The output from this Exclusive-OR circuit will be, in this example, six additional bits, now referred to as Syndrome bits. If the Syndrome bits are all zero, then no errors occurred (Fig. II.15.5). If the two sets of Check bits do not agree, then the Exclusive-OR comparison will be non-zero, (Fig. II.15.6), indicating that an error has occurred. The Syndrome bit pattern resulting indicates the data bit that is in error. Processing circuitry within the 2960, if enabled, will use the Syndrome bit pattern to correct the data error if possible. Figure II.15.7 shows the 2960 parity checking arrangement used to generate Check bits for 16-bit data while Figure II.15.8 shows how the Syndrome bit pattern is decoded to identify errors. Note that in Figure II.15.8 several kinds of errors are possible. If the error is a single-bit error, the EDC chip can correct the error by simply changing the state of the bit. For example, if the Syndrome bits indicate that data bit three is in error and data bit three is currently a one, then the EDC hardware inverts

Figure II.15.3 Check Bit Generation

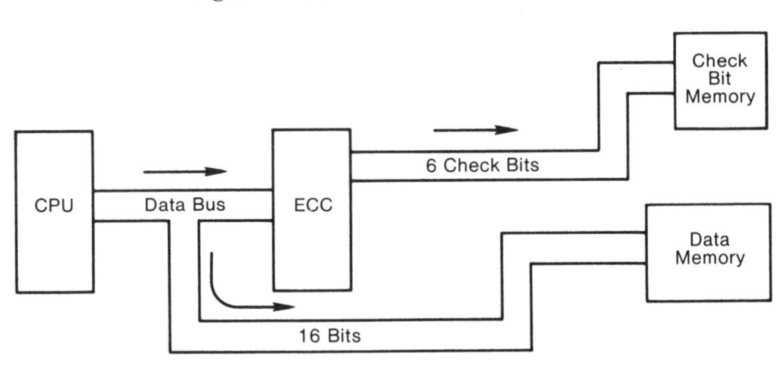

Figure II.15.4 Error Detection/Correction Data Flow

Figure II.15.5 Syndrome Bit Generation
Syndrome bits = all zeros, therefore, no errors.

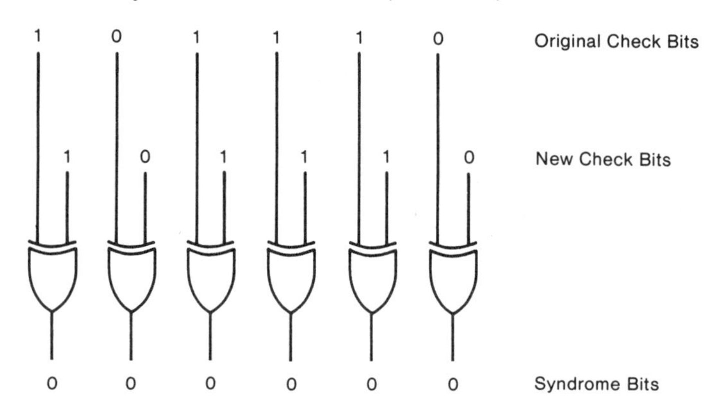

Figure II.15.6 Syndrome Bit Generation
Syndrome bits ≠ all zeros. Syndrome bit pattern indicates the type of error.

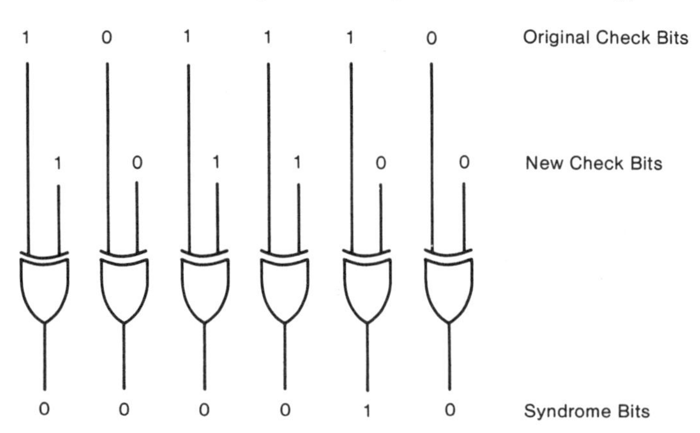

Figure II.15.7 16-bit Modified Hamming Code

Courtesy of Signetics

GENERATED CHECK BITS	PARITY	PARTICIPATING DATA BITS															
		0	1	2	3	4	5	6	7	8	9	10	11	12	13	14	15
CX	Even (XOR)		X	X	X		X			X	X		X			X	
C0	Even (XOR)	X	X	X		X		X		X		X		X			
C1	Odd (XNOR)	X			X	X			X		X	X			X		X
C2	Odd (XNOR)	X	X				X	X	X				X	X	X		
C4	Even (XOR)			X	X	X	X	X	X							X	X
C8	Even (XOR)									X	X	X	X	X	X	X	X

NOTE:
The check bit is generated as either an XOR or XNOR of the eight data bits noted by an "X" in the table.

March 1985

data bit three to zero, which returns it to its original, correct state. The error could just as easily be one of the original Check bits. In this case, the desired data is correct so this error can be ignored (the 2960 can regenerate the Check bits if desired). Finally, two or more errors may have been detected. The EDC hardware cannot identify specific multiple bit errors, so correction is not possible. However, since Syndrome bits are available as chip outputs, this information can be stored for data logging and maintenance purposes. If, for instance, the same memory chip begins to have a rash of single or multiple bit errors, it is most probably an indication of an imminent hard error on that chip.

Figure II.15.8 Syndrome Decode to Bit-in-Error

Courtesy of Signetics

SYNDROME BITS											
			S8	0	1	0	1	0	1	0	1
			S4	0	0	1	1	0	0	1	1
			S2	0	0	0	0	1	1	1	1
SX	S0	S1									
0	0	0		*	**C8**	**C4**	T	**C2**	T	T	M
0	0	1		**C1**	T	T	**15**	T	**13**	**7**	T
0	1	0		**C0**	T	T	M	T	**12**	**6**	T
0	1	1		T	**10**	**4**	T	**0**	T	T	M
1	0	0		**CX**	T	T	**14**	T	**11**	**5**	T
1	0	1		T	**9**	**3**	T	M	T	T	M
1	1	0		T	**8**	**2**	T	**1**	T	T	M
1	1	1		M	T	T	M	T	M	M	T

* — no errors detected
Number — the location of the single bit-in-error
T — two errors detected
M — three or more errors detected

INTERFACING THE EDC TO A MICROPROCESSOR AND MEMORY

Sixteen, bidirectional data lines (Data Bus) provide the means for data transfer between EDC and CPU. Data supplied to the EDC is latched internally in the Data Input Latches under control of the Latch Enable Input line (LE IN). A similar set of latches, the Diagnostic Latches, is also accessible from the Data Bus. Data destined for these latches is synchronized for input by the Diagnostic Latch Enable line (LE DIAG).

A separate set of latches, called the Data Output Latches, stores data before it is transferred to the data bus. The data contained therein may have been modified by the error correction logic. The mode of operation established at the time the Latch Enable Output (LE OUT) is asserted determines the actual contents of the latch. This 16-bit latch can be independently controlled on a byte basis using control lines Output Enable 0 ($\overline{\mathrm{OE}}$ BYTE 0) and Output Enable 1 ($\overline{\mathrm{OE}}$ BYTE 1). When low, the Data Output Latches will be enabled to place their data onto the data bus. When high, the Data Output Latch is placed into a high impedance condition.

Check bits are input to the chip on seven input lines (CB0 - CB6). The Check bits are latched in the same manner as the input data, with control line LE IN.

Check bits or Syndrome bits are output on tri-state lines SC0 through SC6. An output enable line, $\overline{\mathrm{OE}}$ SC (Output Enable Syndrome/Check Bits), controls the tri-state/enable sense of these outputs.

The actual interface to a microprocessor depends on the processor chosen and the method used to implement an EDC system. Some external logic is required to sort out the timings of the various input and output enable lines. The placement of the EDC chip in the memory data path can also vary. The chip can be directly in the data path so that every data transfer is required to pass through the EDC chip. This is easy to set up but tends to slow down data transfers. Although the worst-case time in which data can be corrected is only 65 nsec, this may not be fast enough for some applications. The EDC chip could instead be placed in a configuration where data is monitored in parallel with system data transfers but the EDC intervenes only if a data error occurs. Other applications may dictate the use of multiple EDC subsystems in a computer's memory system.

Typically, an error condition is brought to the CPU's attention through an interrupt. The EDC chip signals errors by bringing $\overline{\mathrm{ERROR}}$, $\overline{\mathrm{MULT\ ERROR}}$ or both low. These lines are used to interrupt the CPU. $\overline{\mathrm{ERROR}}$ is low when one or more errors are detected. $\overline{\mathrm{MULT}}$

$\overline{\text{ERROR}}$ is low if two or more errors are found. The CPU software response to the interrupt is applications dependent.

EDC OPERATING MODES

Several modes of operation, controlled by input pins, are possible with the 2960. Figure II.15.9 shows how the Diagnostic Mode inputs (DIAG MODE 0 and DIAG MODE 1) and the Generate input enable these modes. Figure II.15.10 further defines the modes to be discussed. The Initialize mode (DM1, DM0 = 11, $\overline{\text{GENERATE}}$ = X) is useful at system power-up. In this mode, all bits in the Data Output Latch are forced to zero and the Check bits corresponding to this data combination are generated and made available at the Syndrome/Check Bit outputs. System memory can then be loaded entirely with this pattern, insuring a known starting point.

Normal operation occurs when the Diagnostic Mode bits are both zero. The EDC may now generate Check bits, correct errors, or detect errors. If the Generate input is low, then the EDC chip is placed into Generate mode. In this mode, check bits are generated for data stored in the Input Latch.

Raising the Generate line removes the chip from Generate Mode. The Correct input line determines whether input data is corrected or merely examined for errors. When Correct is high, single-bit errors are corrected by the chip. When Correct is low, errors are only detected. In both cases, the Check bits accompanying the data are compared with

Figure II.15.9 EDC Operating Modes
Courtesy of Signetics

OPERATING MODE	DIAGNOSTIC MODE**		$\overline{\text{GENERATE}}$	
	DM_1	DM_0	0	1
Normal	0	0	Generate	Correct*
Diagnostic Generate	0	1	Diagnostic Generate	Correct*
Diagnostic Correct	1	0	Generate	Diagnostic Correct*
Initialize	1	1	Initialize	Initialize
Pass Thru	When PASS THRU is asserted the Operating Mode is defaulted to the Pass Thru Mode.			

*Correct if the CORRECT Input is HIGH, Detect if the CORRECT Input is LOW.
**In Code ID_{2-0} 001 (ID_2, ID_1, ID_0) DM_1 and DM_0 are taken from the Diagnostic Latch.

Figure II.15.10 Diagnostic Mode Control
Courtesy of Signetics

DIAG $MODE_1$	DIAG $MODE_0$	DIAGNOSTIC MODE SELECTED
0	0	**Non-diagnostic mode.** The EDC functions normally in all modes.
0	1	**Diagnostic Generate.** The contents of the Diagnostic Latch are substituted for the normally generated check bits when in the Generate Mode. The EDC functions normally in the Detect or Correct modes.
1	0	**Diagnostic Detect/Correct.** In the Detect or Correct Mode, the contents of the Diagnostic Latch are substituted for the check bits normally read from the Check Bit Input Latch. The EDC functions normally in the Generate Mode.
1	1	**Initialize.** The outputs of the Data Input Latch are forced to zeroes (and latched upon removal of the Initialize Mode) and the check bits generated correspond to the all-zero data.

newly generated Check bits. The error lines $\overline{\text{ERROR}}$ and $\overline{\text{MULT ERROR}}$ are active and correspond to the error condition detected.

If input PASS THRU is active (high) then the other modes are overridden. Input data and Check bits pass through the chip to their respective outputs untouched.

The two diagnostic modes are discussed in the diagnostics section below.

Configuring the EDC for Data Length

The most common configurations for this EDC chip are 16,- 32, and 64-bit data words. Data sizes other than these can be handled but require additional hardware (latches) and some software overhead. A single EDC chip can handle 16-bit data, which is processed as described above. Data quantities larger than this require the use of more than one EDC chip. The EDC chip or chips are informed of their responsibilities in the system through the information set on the CODE ID 0-2 lines. Figure II.15.11 displays the eight code possibilities for an EDC chip. CODE 000, for instance, sets up the EDC chip for 16-bit operation. The reference 16/22 means that the data word is 16 bits long and that, including the Check bits, 22 bits are actually used. Notice that when the data word is 64 bits long, four EDC chips are necessary. The various codes (100 through 111) inform the individual chips which slice or bytes of the data word they are to process. Also notice that a

Figure II.15.11 Hamming Code and Slice Identification
Courtesy of Signetics

CODE ID_2	CODE ID_1	CODE ID_0	HAMMING CODE AND SLICE SELECTED
0	0	0	Code 16/22
0	0	1	Internal Control Mode
0	1	0	Code 32/39, Bytes 0 and 1
0	1	1	Code 32/39, Bytes 2 and 3
1	0	0	Code 64/72, Bytes 0 and 1
1	0	1	Code 64/72, Bytes 2 and 3
1	1	0	Code 64/72, Bytes 4 and 5
1	1	1	Code 64/72, Bytes 6 and 7

32-bit data word requires two EDC chips. Code 001 (Internal Control Mode) is discussed in the Diagnostics section.

In both the 32-bit and 64-bit configurations, the Check bits and Syndrome bits generated depend on logic existing in multiple EDC chips. The various EDC chips produce partial information that must be combined to create the final, usable Check/Syndrome bits. Figure II.15.12 shows how this is done for the 32-bit case. The low order chip Syndrome/Check bit outputs (SC0 - SC6) are fed to the high order chip Check bit inputs in a straightforward manner. This allows the second chip to know what the first has done. However, the first chip must know what is going on in chip number two, so the second chip Syndrome/Check bit outputs are connected to the Check bit inputs of the first. But, there is a difference this time. The Check bits coming from memory must share the same Check bit input lines of the first (low order) chip with the Syndrome/Check bit information from the second chip. Tying these inputs directly together results in a bus contention problem. Therefore, a tri-state buffer is inserted between the memory Check bit lines and the EDC chip. The Output Enable control line for the Syndrome/Check bit outputs ($\overline{OE}$ SC) is used to switch between the two check bit sources.

The 64-bit configuration (Figure II.15.13) requires Syndrome/Check bit information and obtains it in a fashion similar to that in the 32-bit circuit. Tri-state buffers control where and when the information is presented to the EDC inputs. Notice that additional external logic is required to complete the EDC system. Also notice that the error output lines have been modified as compared to other configurations. The external logic produces additional information to determine if single, double, or three or more errors has occurred (useful for data logging).

Figure II.15.12 32-bit Data Format and I/O

Courtesy of Signetics

Uses Modified Hamming Code 32/39
- 32 data bits
- 7 check bits
- 39 bits in total

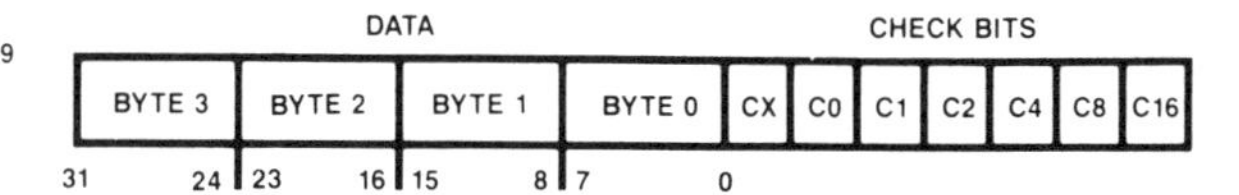

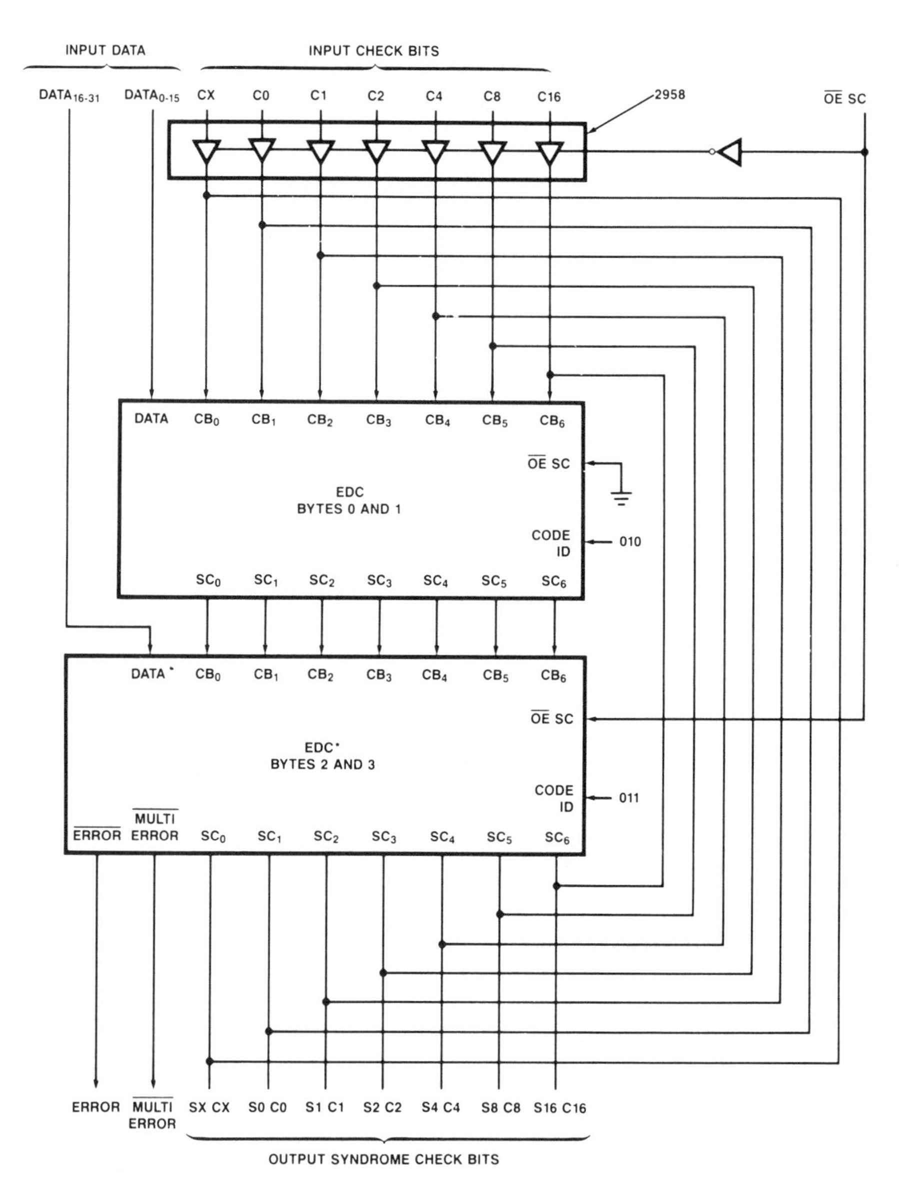

*Check Bit Latch is forced transparent in this CODE ID combination for this slice.

Figure II.15.13 64-bit Data Format and I/O Configuration

Courtesy of Signetics

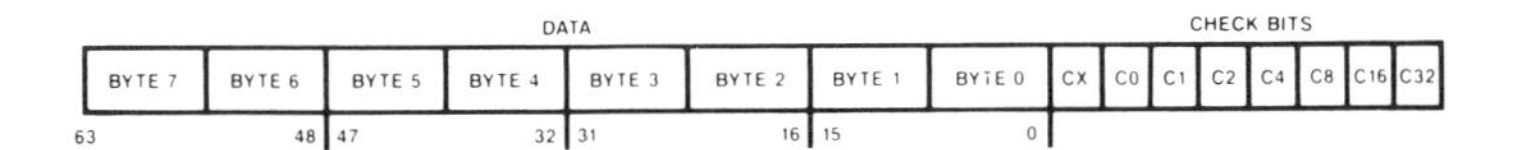

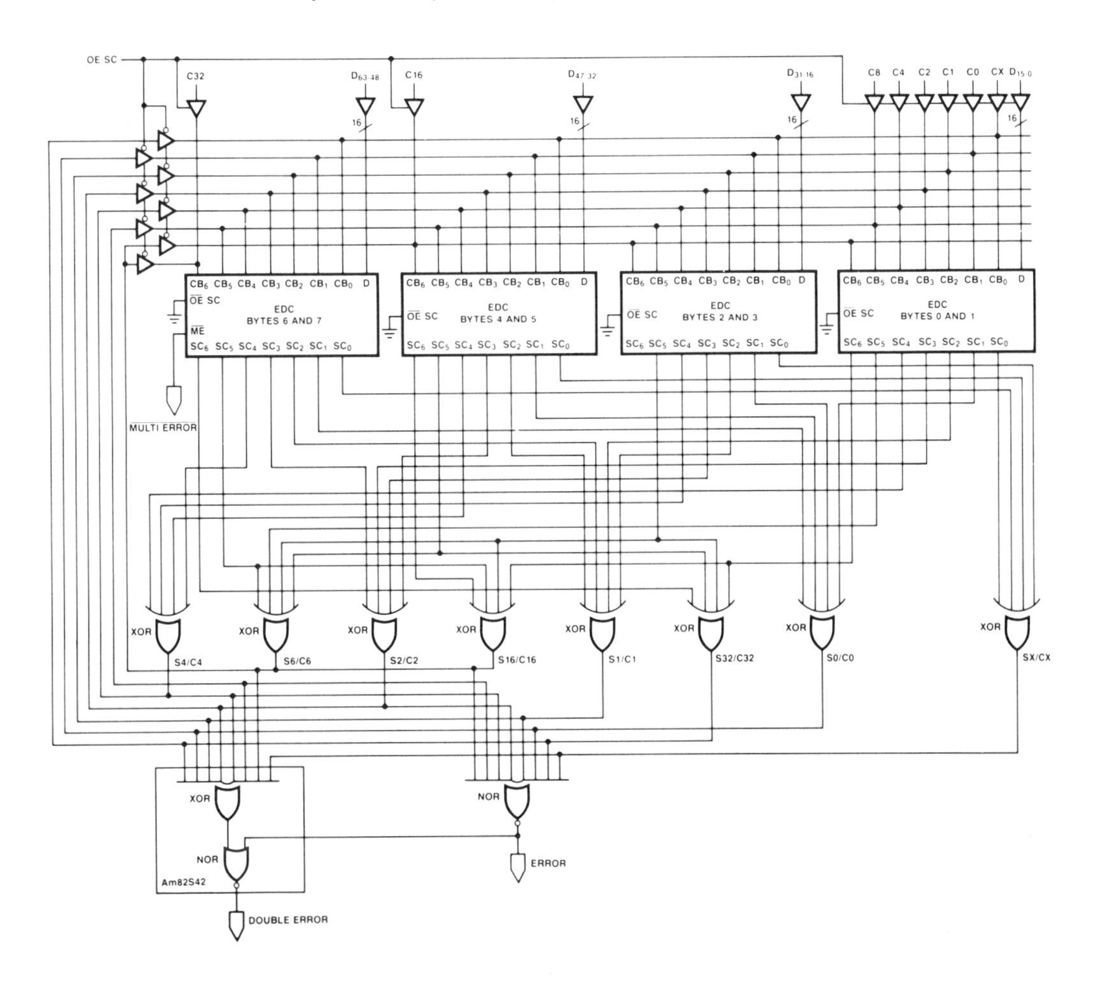

The $\overline{ERROR}$ line is low if one or more errors are detected. The DOUBLE ERROR line is high only if a double bit error is spotted. The $\overline{MULT\ ERROR}$ line (any of the four is valid) is low for all triple, and some double bit errors.

Diagnostics

Built-in diagnostics, selected as shown in Figures II.15.9 and II.15.10, allow the EDC hardware to be tested with software. Sixteen diagnostic latches are loaded with diagnostic check bits for use with CPU test software routines. Naturally, these routines are written to reflect the peculiarities of the EDC system under test.

The Diagnostic Latches can also exhibit another function. When CODE ID 0-2 is set to 001 (Figure II.15.11), the EDC operates under Internal Control Mode. Figure II.15.14 shows the definition of the bits in the Diagnostic Latches. Notice that in addition to the Diagnostic Check bits, the latch also mimics the function of many of the controlling input pins of the 2960. When Internal Control Mode is selected, the EDC reacts to the control signals from this latch rather than from the external control pins, thus providing another useful diagnostic feature.

Figure II.15.14 Diagnostic Latch Loading
Courtesy of Signetics

DATA BIT	INTERNAL FUNCTION
0	Diagnostic Check Bit X
1	Diagnostic Check Bit 0
2	Diagnostic Check Bit 1
3	Diagnostic Check Bit 2
4	Diagnostic Check Bit 4
5	Diagnostic Check Bit 8
6, 7	Don't Care
8	CODE ID 0
9	CODE ID 1
10	CODE ID 2
11	DIAG MODE 0
12	DIAG MODE 1
13	CORRECT
14	PASS THRU
15	Don't Care

A full system error correction and detection implementation cannot be designed as an afterthought because of the highly complex nature of both EDC and memory systems. A high performance, efficient system is designed only after careful thought is given to the system hardware and software requirements.

SECTION III

Integrated Circuit Reference

III.1

Directory of Peripheral Chip Part Numbers

CHIP TYPE	PART NUMBER	MANUFACTURER
A/D Converter	CDP68HC68A1	RCA
A/D Converter	Si520	Siliconix
A/D Converter	Si8601	Siliconix
A/D Converter	Si8602	Siliconix
A/D Converter	ADC-910	Precision Monolithics
A/D Converter	ADC-8208	Precision Monolithics
A/D Converter	TSC7109	Teledyne
A/D Converter	TSC8700	Teledyne
A/D Converter	TSC8701	Teledyne
A/D Converter	TSC8702	Teledyne
A/D Converter	ICL7109	Intersil
A/D Converter	AD573	Analog Devices
A/D Converter	AD574A	Analog Devices
A/D Converter	AD673	Analog Devices
A/D Converter	AD7552	Analog Devices
A/D Converter	AD7571	Analog Devices
A/D Converter	AD7574	Analog Devices
A/D Converter	AD7581	Analog Devices
A/D Converter	ADC0801	Intersil
A/D Converter	ADC0802	Intersil
A/D Converter	ADC0803	Intersil
A/D Converter	ADC0804	Intersil
Bus Arbiter	8289	Intel
Bus Arbiter	8289-1	Intel
Bus Arbiter	82289	Intel

PINS	MICROCOMPATIBILITY	COMMENTS
16	CDP1800, 6805, 8048/49/51/85/88, NSC800, Z80, 6502	Serial 10-Bit A/D Converter
28	8080/85, Z80, SC/MP, 6800	8-Channel 8-Bit A/D
28	8080/85, Z80, SC/MP, 6800	8-Channel 8-Bit A/D
28	8080/85, Z80, SC/MP, 6800	8 Channel 8-Bit A/D
28	6502, 68000	10-Bit A/D
18		8-Bit A/D
40-DIP, 60-Flat Package	8080, 8085	
24	8080	8-Bit A/D
24	8080	10-Bit A/D
24	8080	12-Bit A/D
40	8008, 8048, 8080/85, 6800, IM6100	12-Bit A/D
20	6502, 8085	10-Bit A/D
28	8080/85, 6502, 6800	12-Bit A/D
20		8-Bit A/D
40	6800, 8085	12-Bit Plus Sign A/D
28	6502, 68000, 8085	10-Bit Plus Sign A/D
18		8-Bit A/D
28	8080/85, Z80, 6800	8 Channel 8-Bit A/D
20	8048, 8080/85, Z80, 6800	8-Bit A/D
20	8048, 8080/85, Z80, 6800	8-Bit A/D
20	8048, 8080/85, Z80, 6800	8-Bit A/D
20	8048, 8080/85, Z80, 6800	8-Bit A/D
20	8086/88, 80186	8 MHz
20	8086/88, 80186	10 MHz
20	80286	

CHIP TYPE	PART NUMBER	MANUFACTURER
Clock Generator	8224	Intel
Clock Generator	8284A	Intel
Clock Generator	8284A-1	Intel
Clock Generator	82C84A	Intel
Clock Generator	82C84A-5	Intel
Clock Generator	82284	Intel
Clock Generator	82384	Intel
Clock Generator	Z8581	Zilog
Communications	2651, COM2651, INS2651	Signetics, Standard Microsystems, National
Communications	2681	Signetics
Communications	2681	Signetics
Communications	2641	Signetics
Communications	MK68564	Mostek
Communications	SCN68562C4N48	Signetics
Communications	SCN68681C1N40	Signetics
Communications	2652, COM5025, SND5025	Signetics, Standard Microsystems, Solid State Scientific
Communications	2652-1	Signetics
Communications	2661-1, AY2661, COM2661-1, SY2661	Signetics, General Inst., Standard Microsystems, Synerte
Communications	2661-2, COM2661-2	Signetics, Standard Microsystems
Communications	2661-3, COM2661-3	Signetics, Standard Microsystems
Communications	82586	Intel
Communications	82588	Intel
Communications	82501	Intel
Communications	82502	Intel
Communications	8251A, 8251A, COM8251A, INS8251A, uPD8251A	AMD, Intel, Standard Microsystems, National, NEC

PINS	MICROCOMPATIBILITY	COMMENTS
16	8080	
18	8086/88	5 MHz and 8 MHz
18	8086/88	10 MHz
18-Dip, 20-PLCC	80C86/88	8 MHz
18-DIP, 20-PLCC	80C86/88	5 MHz
18	80286	
18	80386	
18	Z80, Z8000	Clock Generator and Controller
28		Programmable Communications Interface
40		
28/24		
48	68000	Serial I/O Controller
	68000	Serial Communications Controller
	68000	Dual UART
40	68000	Multi-Protocol Communications Controller
	68000	Multi-Protocol Communications Controller
28	68000	Enhanced Programmable Communication Interface
28	68000	Enhanced Programmable Communication Interface
28	68000	Enhanced Programmable Communication Interface
48		Local Area Network Coprocessor
28	80186/188	Local Area Network Controller
20		Ethernet Serial Interface
16		Ethernet Transceiver
28	8048, 8080, 8085, 8086/88	USART

CHIP TYPE	PART NUMBER	MANUFACTURER
Communications	8273	Intel
Communications	8273-4	Intel
Communications	8274	Intel
Communications	82530	Intel
Communications	82530-6	Intel
Communications	8291A	Intel
Communications	8292	Intel
Communications	INS1671, COM1671, UC1671	National, Standard MicroSystems, Western Digital
Communications	COM7210, uPD7210	Standard Microsystems, NEC
Communications	COM9026	Standard Microsystems
Communications	AY-3-1013, HD6402, H1854A, IM6402, CDP1854	GI, Harris, Hughes, Intersil, RCA, Western Digital
Communications	H1854AC, CDP1854AC	Hughes, RCA
Communications	CDP6402	RCA
Communications	CDP6402C	RCA
Communications	CDP65C51	RCA
Communications	CDP6853	RCA
Communications	S2350	Gould AMI
Communications	Z8440	Zilog
Communications	Z8470	Zilog
Communications	Z8030	Zilog
Communications	Z8031	Zilog
Communications	Z8530	Zilog
Communications	Z8531	Zilog
Communications	IM6402	Intersil
Communications	IM6403	Intersil

PINS	MICROCOMPATIBILITY	COMMENTS
40	8048, 8080, 8085, 8086/88, 80186/188	HDLC/SDLC Controller
40	8048, 8080, 8085, 8086/88, 80186/188	HDLC/SDLC Controller
40	8048, 8051, 8085, 8086/88, 80186/188, 8089	Serial Communications Controller
40	8051, 8086/88, 80186/188	Serial Communications Controller
40	8051, 8086/88, 80186/188	Serial Communications Controller
40	8048/49, 8051, 8080/85, 8086/88	GPIB Talker/Listener
40	8048/49, 8051, 8080/85, 8086/88	GPIB Controller
40		USART
40	8080/85, 8086	GPIB Interface Controller
40		Local Area Network Controller
40	CDP1800, 6805, 8048/49/51/85/88, NSC800, Z80, 6502	UART
40	CDP1800, 6805, 8048/49/51/85/88, NSC800, Z80, 6502	UART
40	CDP1800, 6805, 8048/49/51/85/88, NSC800, Z80, 6502	UART
40	CDP1800, 6805, 8048/49/51/85/88, NSC800, Z80, 6502	UART
28	CDP1800, Z80, 6502	Asynchronous Communications Interface Adapter
28	6805, 8048/49/51/85/88, NSC800	Asynchronous Communications Interface Adapter
40	6800	USRT
40	Z80	Serial I/O Controller
40	Z80	Dual UART
40	Z8000	Serial Communcations Controller
40	Z8000	Asynchronous Serial Communications Controller
40	Z80	Serial Communications Controller
40	Z80	Asynchronous Serial Controller
40	IM6100	UART
40	IM6100	UART

CHIP TYPE	PART NUMBER	MANUFACTURER
Communications	TMS38010	Texas Instruments
Communications	TMS38020	Texas Instruments
Communications	TMS38030	Texas Instruments
Communications	TMS38051	Texas Instruments
Communications	TMS38052	Texas Instruments
Communications	TMS9914A	Texas Instruments
Communications	S1602, 5303	Gould AMI, National
Communications	S2350	Gould AMI
Communications	S6551/S6551A, 6551, SY6551	Gould AMI, Rockwell, Synertek
Communications	S6850, MC6850, F6850	Gould AMI, Motorola, Fairchild
Communications	S68A50	Gould AMI
Communications	S68B50	Gould AMI
Communications	S6852, MC6852, F6852	Gould AMI, Motorola, Fairchild
Communications	S68A52	Gould AMI
Communications	S68B52	Gould AMI
Communications	S6854, MC6854, F6854	Gould AMI, Motorola, Fairchild
Communications	S68A54	Gould AMI
Communciations	S68B54	Gould AMI
Communications	S9902	Gould AMI
D/A Converter	Si8021	Siliconix
D/A Converter	Si8020	Siliconix
D/A Converter	XR-9201	Exar
D/A Converter	DAC-888	Precision Monolithics
D/A Converter	DAC-8012	Precision Monolithics
D/A Converter	DAC-8408	Precision Monolithics

PINS	MICROCOMPATIBILITY	COMMENTS
40	MPU800, NSC800	I/O, ROM
40	CDP1800, Z80, 6502	I/O Ports
40	CDP1800, Z80, 6502	I/O Ports
24	DCP1800, 6805, NSC800, Z80, 6502, 8048/51/49/85/88	I/O Port
24	CDP1800, 6805, NSC800, Z80, 6502, 8048/51/49/85/88	I/O Ports
22	CDP1800, 6805, 8048/49/51/85/88, NSC800, Z80, 6502	8-Bit Input Port
22	CDP1800, 6805, 8048/49/51/85/88, NSC800, Z80, 6502	8-Bit Input Port
22	CDP1800, 6805, 8048/49/51/85/88, NSC800, Z80, 6502	8-Bit Output Port
40	6805, 8048/49/51/85/88, NSC800	Parallel Interface
40	Z80	Parallel I/O Controller
40	IM6100	Parallel I/O
32		Configurable I/O
40	6800	Peripheral Interface Adapter
40	S68A00	Peripheral Interface Adapter
40	S68B00	Peripheral Interface Adapter
40	6800	ROM-I/O-Timer
28	8080, 8085, 8086/88	
28	8080, 8085, 8086/88	
28	8080, 8085, 8086/88	
28	8080, 8085, 8086/88, 80C86/88	
28	CDP1800	Interrupt Controller
28	CDP1800	Interrupt Controller
40	TI32000	Interrupt Control Unit
40	8-bit processors	2 MHz
40	8-bit processors	3.125 MHz

CHIP TYPE	PART NUMBER	MANUFACTURER
D/A Converter	PM-7524	Precision Monolithics
D/A Converter	PM-7528	Precision Monolithics
D/A Converter	PM-7545	Precision Monolithics
D/A Converter	PM-7645	Precision Monolithics
D/A Converter	AD390	Analog Devices
D/A Converter	AD558	Analog Devices
D/A Converter	AD567	Analog Devices
D/A Converter	D667	Analog Devices
D/A Converter	AD7534	Analog Devices
D/A Converter	AD7548	Analog Devices
Data Encryption	8294A	Intel
Data Encryption	Z8068	Zilog
Data Encryption	TMS7500	Texas Instruments
Data Encryption	TMS75C00	Texas Instruments
Disk Controller	82062	Intel
Disk Controller	82064	Intel
Disk Controller	SCB68454C6N48	Signetics
Disk Controller	FDC7265, uPD7265	Standard Microsystems, NEC
Disk Controller	FDC1791-02, FD1791-02, MB8876, SAB-1791	Standard Microsystems, Western Digital, Fujitsu, Siemens, Synertec
Disk Controller	FDC1792-02, FD1792-02	Standard Microsystems, Western Digital
Disk Controller	FDC1793-02, FD1793-02, MB8877, SAB-1793	Standard Microsystems, Western Digital, Fujitsu, Siemens, Synertec
Disk Controller	FDC1794-02, FD1794-02	Standard Microsystems, Western Digital
Disk Controller	FDC1795-02, FD1795-02, SAB-1795	Standard Microsystems, Western Digital, Siemens
Disk Controller	FDC1797-02, FD1797-02, SAB-1797	Standard Microsystems, Western Digital, Siemens
Disk Controller	FDC9266	Standard Microsystems

PINS	MICROCOMPATIBILITY	COMMENTS
16	6800, 8080/85, Z80	8-Bit D/A
20	6800, 8080/85, Z80	Dual 8-Bit D/A
20	6800, 8080, Z80	12-Bit D/A
20	6800, 8080, Z80	12-Bit D/A
28		Quad 12-Bit D/A
16		8-Bit D/A
28		12-Bit D/A
28		12-Bit D/A
20		14-Bit D/A
20	6800, 8085, 6502, Z80	12-Bit D/A
40	8041, 8051, 8080/85, 8086/88	
40	Z8000	Data Ciphering Processor
40	TMS9995	Data Encryption Device
40	TMS9995	Data Encryption Device
40		For Winchester Drives
40		For Winchester Drives
	68000	Multiple Disk Controller
40		Micro-Floppy Disk Controller
40		Floppy Disk Controller/ Formatter
40		Floppy Disk Controller/ Formatter
40		Floppy Disk Controller/ Formatter
40		Floppy Disk Controller/ Formatter
40		Floppy Disk Controller/ Formatter
40		Floppy Disk Controller/ Formatter
40		Enhanced Floppy Disk Controller

CHIP TYPE	PART NUMBER	MANUFACTURER
Disk Controller	FDC9791	Standard Microsystems
Disk Controller	FDC9793	Standard Microsystems
Disk Controller	FDC9795	Standard Microsystems
Disk Controller	FDC9797	Standard Microsystems
Disk Controller	HDC7261A, uPD7261A	Standard Microsystems, NEC
Disk Controller	HDC9224	Standard Microsystems
Disk Controller	TMS2791	Texas Instruments
Disk Controller	TMS2793	Texas Instruments
Disk Controller	TMS2795	Texas Instruments
Disk Controller	TMS2797	Texas Instruments
Disk Controllers	765A, 8272A, FDC765A, uPD765	Rockwell, Intel, Standard Microsystems, NEC
DMA Controller	8237A	Advanced Micro Devices, Intel
DMA Controller	8237A-4	Intel
DMA Controller	8237A-5	Intel
DMA Controller	82C37A-5	Intel
DMA Controller	8257	Intel
DMA Controller	8257-5	Intel
DMA Controller	Z8410	Zilog
DMA Controller	Z8016, Z8016	Advanced Micro Devices, Zilog
DMA Controller	Am2940	Advanced Micro Devices
DMA Controller	Am2942	Advanced Micro Devices
DMA Controller	82258	Intel
DMA Controller	Am9516A	Advanced Micro Devices
DMA Controller	Am9517A	Advanced Micro Devices
DMA Controller	Am9517A-4	Advanced Micro Devices

PINS	MICROCOMPATIBILITY	COMMENTS
40		Floppy Disk Controller/ Formatter
40		Floppy Disk Controller/ Formatter
40		Floppy Disk Controller/ Formatter
40		Floppy Disk Controller/ Formatter
40		Hard Disk Controller
40		Universal Disk Controller
40		Floppy Disk Formatter/ Controller
40		Floppy Disk Formatter/ Controller
40		Floppy Disk Formatter/ Controller
40		Floppy Disk Formatter/ Controller
40		Floppy Disk Controller
40	8085, 8086/88	3 MHz
40	8085, 8086/88	4 MHz
40	8085, 8086/88	5 MHz
40-DIP; 44-PLCC	8085, 8086/88	5 MHz
40	8085	
40	8085	
40	Z80	
48	Z8000	DMA Access Controller
28-DIP, 28-LCC		DMA Address Generator only
22-DIP, 28-LCC		DMA Address Generator, Timer/Counter only
68-LCC	8086/88, 80186/188, 80286	6 and 8 MHz
48	8086/88, 80186/188, 80286, 68000	
40	8085, 8086/88, 80186/188, 80286, 68000	3 MHz
40	8085, 8086/88, 80186/188, 80286, 68000	4 MHz

CHIP TYPE	PART NUMBER	MANUFACTURER
DMA Controller	Am9517A-5	Advanced Micro Devices
DMA Controller	SCB68430CA148	Signetics
Dynamic Ram Controller	8202A	Intel
Dynamic Ram Controller	8202A-1	Intel
Dynamic Ram Controller	8202A-3	Intel
Dynamic Ram Controller	8203	Intel
Dynamic Ram Controller	8203-1	Intel
Dynamic Ram Controller	8203-3	Intel
Dynamic RAM Controller	8207	Intel
Dynamic RAM Controller	8208	Intel
Dynamic RAM Controller	82C08	Intel
Error Correction	8206	Intel
Error Correction	8206-1	Intel
Error Correction	8206-2	Intel
Error Detection	2960	Signetics
Error Detection	Z8065	Zilog
I/O	8755A	Intel
I/O	8755A-2	Intel
I/O	8255A	Intel
I/O	8255A-5	Intel
I/O	82C55A	Intel
I/O	MPU810A, NSC810A	Standard Microsystems, National
I/O	MPU810A-1	Standard Microsystems
I/O	MPU810A-4	Standard Microsystems
I/O	MPU830, NSC830	Standard Microsystems, National

PINS	MICROCOMPATIBILITY	COMMENTS
40	8085, 8086/88, 80186/188, 80286, 68000	5 MHz
40	8080, 8085, 8086/88	2117, 2118 RAMs
40	8080, 8085, 8086/88	2117, 2118 RAMs
40	8080, 8085, 8086/88	2117, 2118 RAMs
40	8080, 8085, 8086/88	2164, 2118, 2117 RAMs
40	8080, 8085, 8086/88	2164, 2118, 2117 RAMs
40	8080, 8085, 8086/88	2164, 2118, 2117 RAMs
68-LCC	8086/88, 80186/188, 80286	Dual-Port
48	8086/88, 80186/188, 80286	Single Port Version of 8207
48	8086/88, 80186/188, 80286	Single Port Version of 8207
68-LCC	8086/88, 80186/188	42 nsec Detection
68-LCC	8086/88, 80186/188	35 nsec Detection
68-LCC	8086/88, 80186/188	57 nsec Detection
48		
40	Z8000	Burst Error Processor
40	8085, 8088	2K × 8 EPROM with I/O ports
40	8085, 8088	2K × 8 EPROM with I/O ports
40	8085	
40	8085	
40-DIP, 44-PLCC	Most Processors	
40	MPU800, NSC800	RAM, I/O, Timer
40	MPU800, NSC800	RAM, I/O, Timer
40	MPU800, NSC800	RAM, I/O, Timer
40	MPU800, NSC800	I/O, ROM

CHIP TYPE	PART NUMBER	MANUFACTURER
I/O	MPU831, NSC831	Standard Microsystems, National
I/O	CDP1851	RCA
I/O	CDP1851C	RCA
I/O	MK386K, H1852, 8212, INSB212, CDP1852	Fairchild, Hughes, Intel, National, RCA
I/O	H1852C, CDP1852C	Hughes, RCA
I/O	CDP1872C	RCA
I/O	CDP1874C	RCA
I/O	CDP1875C	RCA
I/O	CDP6823	RCA
I/O	Z8420	Zilog
I/O	IM6103	Intersil
I/O	TMS70XX	Texas Instruments
I/O	S6821, MC6821, F6821	Gould AMI, Motorola, Fairchild
I/O	S68A21	Gould AMI
I/O	S68B21	Gould AMI
I/O	S6846, MC6846, F6846	Gould AMI, Motorola, Fairchild
Interrupt Controller	8259A	Intel
Interrupt Controller	8259A-2	Intel
Interrupt Controller	8259A-8	Intel
Interrupt Controller	82C59A-2	Intel
Interrupt Controller	CDP1877	RCA
Interrupt Controller	CDP1877C	RCA
Interrupt Controller	TI32202-2	National Semiconductor, Texas Instruments
Keyboard/Display	8279	Intel
Keyboard/Display	8279-5	Intel

PINS	MICROCOMPATIBILITY	COMMENTS
40	MPU800, NSC800	I/O, ROM
40	CDP1800, Z80, 6502	I/O Ports
40	CDP1800, Z80, 6502	I/O Ports
24	DCP1800, 6805, NSC800, Z80, 6502, 8048/51/49/85/88	I/O Port
24	CDP1800, 6805, NSC800, Z80, 6502, 8048/51/49/85/88	I/O Ports
22	CDP1800, 6805, 8048/49/51/85/88, NSC800, Z80, 6502	8-Bit Input Port
22	CDP1800, 6805, 8048/49/51/85/88, NSC800, Z80, 6502	8-Bit Input Port
22	CDP1800, 6805, 8048/49/51/85/88, NSC800, Z80, 6502	8-Bit Output Port
40	6805, 8048/49/51/85/88, NSC800	Parallel Interface
40	Z80	Parallel I/O Controller
40	IM6100	Parallel I/O
32		Configurable I/O
40	6800	Peripheral Interface Adapter
40	S68A00	Peripheral Interface Adapter
40	S68B00	Peripheral Interface Adapter
40	6800	ROM-I/O-Timer
28	8080, 8085, 8086/88	
28	8080, 8085, 8086/88	
28	8080, 8085, 8086/88	
28	8080, 8085, 8086/88, 80C86/88	
28	CDP1800	Interrupt Controller
28	CDP1800	Interrupt Controller
40	TI32000	Interrupt Control Unit
40	8-bit processors	2 MHz
40	8-bit processors	3.125 MHz

CHIP TYPE	PART NUMBER	MANUFACTURER
Keyboard/Display	KR3600, AY-5-3600	Standard Microsystems, General Instruments
Keyboard/Display	CDP1871A	RCA
Keyboard/Display	CDP1871AC	RCA
Memory Controller	IM6102	Intersil
Memory Controller	TI32082W-2	Texas Instruments
Memory Controllers	Z8003	Zilog
Memory Controllers	Z8004	Zilog
Memory Controllers	Z8010	Zilog
Memory Controllers	Z8015	Zilog
Memory Controllers	TMS4500A	Texas Instruments
Other	8256AH	Intel
Other	2671	Signetics
Other	MK68901	SGS-Thomson
Other	Z8038	Zilog
Other	Z8060	Zilog
Other	Z8090/4	Zilog
Other	Z8590/4	Zilog
Other	IM6101	Intersil
Other	TMS9650	Texas Instruments
Other	S9901	Gould AMI
Other	S9901-4	Gould AMI
Real Time Clock	CDP1879	RCA
Real Time Clock	CDP1879C-1	RCA
Real Time Clock	CDP68HC68T1	RCA
Real Time Clock	CDP6818	RCA

PINS	MICROCOMPATIBILITY	COMMENTS
40		Keyboard Encoder ROM
40	CDP1800, 6805, 8048/49/51/85/88, NSC800, Z80, 6502	Keyboard Encoder
40	CDP1800, 6805, 8048/49/51/85/88, NSC800, Z80, 6502	Keyboard Encoder
40	IM6100	DMA Controller/Timer
48	TI32000	Memory Management Unit
48	Z8000	Virtual Memory processing Unit
40	Z8000	Virtual Memory processing Unit
48	Z8000	Memory Management Unit
64	Z8000	Paged Memory Management Unit
40		Dynamic RAM Controller
40	8086/88, 80186/188, 8051	Timer, Interrupt Controller, Clock, Communications Keyboard and Communications Controller
48	68000	USART, Interrupt Controller, I/O Ports, Timer
40	Z8000	FIFO I/O Interface
28	Z8000	FIFO Buffer and Expander
40	Z8000	Universal Peripheral Controller
40	Z80	Universal Peripheral Controller
40	IM6100	Programmable Interface Element
40	TMS99000, 8088	Multiprocessor Interface
40	S9900	I/O Ports, Interrupts, Real Time Clock
40	S9900	I/O Ports, Interrupts, Real Time Clock
24	CDP1800, Z80, 6502	
24	CDP1800, Z80, 6502	
16	CDP1800, 6805, 8048/49/51/85/88, NSC800, Z80, 6502	Real Time Clock with RAM
24	CDP1800, 6805, 8048/49/51/85/88, NSC800, Z80, 6502	Real Time Clock with RAM

CHIP TYPE	PART NUMBER	MANUFACTURER
Timer	SCN68230C8148	Signetics
Timer	CDP1878	RCA
Timer	CDP1878C	RCA
Timer	CDP6848	RCA
Timer	Z8430	Zilog
Timer	Z8036	Zilog
Timer	Z8536	Zilog
Timer	TI32201-2	Texas Instruments
Timer	S6840, MC6840, F6840	Gould AMI, Motorola, Fairchild
Timer	S68A40	Gould AMI
Timer	S68B40	Gould AMI
Timers	8253	Intel
Timers	8253-5	Intel
Timers	8254	Intel
Timers	8254-5	Intel
Timers	8254-2	Intel
Timers	82C54-2	Intel
Timers	82C54	Intel
Video Controller	8276H	Intel
Video Controller	8276H-2	Intel
Video Controller	82720, CRT7220, uPD7220	Intel, Standard Microsystems, NEC
Video Controller	CRT220A, uPD7220A	Standard Microsystems, NEC
Video Controller	82730	Intel
Video Controller	2670	Signetics
Video Controller	2672	Signetics

PINS	MICROCOMPATIBILITY	COMMENTS
	68000	
28	CDP1800, Z80, 6502	Dual Counter-Timer
28	CDP1800, Z80, 6502	Dual Counter-Timer
28	6805, 8048/49/51/85/88, NSC800	Dual Counter-Timer
28	Z80	Counter/Timer
40	Z8000	Counter/Timer/Parallel I/O
40	Z80	Counter/Timer/Parallel I/O
24	TI32000	Timing Control Unit
28	6800	Programmable Timer
28	6800	Programmable Timer
28	6800	Programmable Timer
24	8085	
24	8085	
24	Most Processors	8 MHz
24	Most Processors	5 MHz
24	Most Processors	10 MHz
24-DIP, 28 PLCC	Most Processors	8 MHz
24-DIP, 28 PLCC	Most Processors	10 MHz
40	8051, 8085, 8086/88	2 MHz
40	8051, 8085, 8086/88	3 MHz
40		Graphics Display Controller
40		Graphics Display Controller
68-LCC		Text Coprocessor
		Character Generator
		Timing Controller

CHIP TYPE	PART NUMBER	MANUFACTURER
Video Controller	2673A	Signetics
Video Controller	2674	Signetics
Video Controller	2675	Signetics
Video Controller	2677	Signetics
Video Controller	2636	Signetics
Video Controller	2637	Signetics
Video Controller	CDP1869C	RCA
Video Controller	CDP1870C	RCA
Video Controller	VSC	Texas Instruments
Video Controller	TMS34070	Texas Instruments
Video Controller	S68045, MC6845	Gould AMI, Motorola
Video Controller	S68A045	Gould AMI
Video Controller	S68B045	Gould AMI
Video Controllers	8275H, CRT5037, MK3807, SND5037, TMS9927	Intel, Standard Microsystems, SGS-Thomson, Solid State Scientific, Texas Instruments
Video Controllers	8275H-2	Intel
Video Controllers	CRT5027	Standard Microsystems
Video Controllers	CRT5057	Standard Microsystems
Video Controllers	2609, CRT7004, MC6570	Signetics, Standard Microsystems, Motorola
Video Controllers	CRT8002, SND8002	Standard Microsystems, Solid State Scientific
Video Controllers	CRT9007A	Standard Microsystems
Video Controllers	CRT9007B	Standard Microsystems
Video Controllers	CRT9007C	Standard Microsystems
Video Controllers	CRT9028	Standard Microsystems
Video Controllers	CRT9128	Standard Microsystems

PINS	MICROCOMPATIBILITY	COMMENTS
		Attributes Controller, several versions
		Color/Mono Attributes Controller
		Attributes Controller
		Video Interface
		Video Interface
40	CDP1802, CDP1804	Address and Sound Generator used with CDP1870C
40	CDP1802, CDP1804	Color Video Generator used with CDP1869C
68-PLCC		Video Systems Controller
22		Video Palette
40	6800	CRT Controller
40	6800	CRT Controller
40	6800	CRT Controller
40	8051, 8085, 8086/88	2 MHZ CRT Controller
40	8051, 8085, 8086/88	3 MHZ
40		CRT Controller
40		CRT Controller
24		Dot Matrix Character Generator
28		
		Video Display Attributes Controller
40		Video Processor and Controller
40		Video Processor and Controller
40		Video Processor and Controller
40	8051, 8085, 8086, Z80	Video Terminal Logic Controller
40	8051, 8085, 8086 Z80	Video Terminal Logic Controller

III.2

Byte-Size Microprocessor and Microcontroller Reference Sheet

1-BIT PROCESSORS

14500B

2-BIT PROCESSORS

3000

4-BIT PROCESSORS

402
402M
402L
410L
411L
420
420L
420C
444L
445L
1000C
1070
1100
1200C
1270
1300
2000A
2150A
2200A
2300
2901
2901A
2901B
2901C
2903
2903A
4004
5781
5782
5799
6701
10800
29203
57140

4-Bit Processors (*cont.*)

54LS481
74LS481
74S481
141000
141200

8-BIT PROCESSORS

F8
Z8
Z80
Z80A
Z80B
Z80L
1650
1802A
1802AC
1802BC
1804AC
1804PCE
1805AC
1806AC
Z800
3870
38E70
38P70
3870
3880
6502
6502C
65C02
6800
68A00
68B00
6801
6802
6803
68HCO4P2
68HCO4P3
6805

8-Bit Processors (*cont.*)

6805E2
6805E3
6805F2
6805G2
68HCO5C4
68HC05D2
6808
7000
7001
7002
7020
7040
7041
7042
70C00
70C20
70C40
70P161
7742
LP8000
8008
8031
8031AH
80C31BH
80C31BH-1
80C31BH-2
8032AH
8035AHL
8035H
8039AHL
8039H
80C39-7
8040AHL
8048AH
8049AH
80C49-7
8050AH
8051
8051AH
80C51BH
80C51BH-1
80C51BH-2
8052AH
8052AH-BASIC
8080A
8080A-1
8080A-2
8085AH
8085AH-1
8085AH-2
80C252
8X305
83C252
8748H
8749H
8751H
8751H-12
8751H-88
87C51
8752A
9085A
46800
46802
87C252
100220
146805

12-BIT PROCESSORS

6100

16-BIT PROCESSORS

V20
V30
mN601
mN602
1600
1610
SX2000
6809
ZS8001

16-Bit Processors (*cont.*)

Z8003
8086
80C86
80C86-2
8094
8095
8096
8097
iAPX 88/10
80C88
80C88-2
80186
80188
8089
9008
9440
9445
9900
9940
9980
9981
9985
16000
29116
68000
68008
68010
68012
68200
80286-6
80286-8
80286-10
99000

32-BIT PROCESSORS

32032D
32032T
32C032
32016T
Z8002
Z8004
8394
8396
8397
8395
8794
8796
8795
8797
3208T
32132
32310
32332
32382
68020
68030
Z80000
80386
Am29000

III.3

Bibliography of Useful Reference Material

Advanced Micro Devices
MOS Microprocessors and Peripherals Data Book

Advanced Micro Devices
Bipolar Microprocessor Logic and Interface Data Book

Advanced Micro Devices
Am9519A Universal Interrupt Controller Technical Manual

Advanced Micro Devices
Programmable Array Logic Handbook

Analog Devices
Data Book

Gould - AMI Semiconductors
MOS Products Catalog

Hughes
CMOS Databook

Intel
Microsystem Components Handbook, Vol. 1 and 2

Intel
Memory Components Handbook

Intel
Microcontroller Handbook

Intel
Microcommunications Handbook

Intersil
Intersil Data Book

Mostek Corporation
Microcomputer Components Technical Manual

Precision Monolithics, Inc.
Linear and Conversion Products Data Book

Raytheon, Semiconductor Division
Product Selection Guide

RCA
CMOS Microprocessors, Memories and Peripherals Data Book

Signetics
Linear LSI Full Product Guide

Standard Microsystems Corporation
Data Catalog

Teledyne Semiconductor
Data Acquisition Handbook

Texas Instruments
The TTL Data Book

Texas Instruments
MOS Memory Data Book

Texas Instruments
TI32000 Family Data Manual

Texas Instruments
TMS7000 Family Data Manual

Zilog
Component Products Profile

III.4

Integrated Circuit Manufacturers

Advanced Micro Devices
901 Thompson Pl.
Sunnyvale, CA 94086
(408-732-2400)

Analog Devices, Inc.
P.O. Box 280
Norwood, MA 02062
(617-329-4700)

Fujitsu Microelectronics, Inc.
3320 Scott Boulevard
Santa Clara, CA 95051

GE, Intersil
10600 Ridgeview Ct.
Cupertino, CA 95014
(408-996-5000)

Gould AMI Semiconductors
3800 Homestead Road
Santa Clara, CA 95051
(408-246-0330)

Hughes Aircraft Co.
Microelectronics Products Division
500 Superior Avenue
Newport Beach, CA 92675
(714-548-0671)

Hyundai Electronics Industries Co., Ltd.
4401 Great America Parkway
Santa Clara, CA 95054

Integrated Device Technology, Inc.
3236 Scott Boulevard
P.O. Box 58015
Santa Clara, CA 95052

Intel Corp.
3065 Bowers Avenue
Santa Clara, CA 95051
(408-246-7501)

Maxim Integrated Products
510 N. Pastoria Avenue
Sunnyvale, CA 94086

Motorola Semiconductor
Products, Inc.
P.O. Box 20912
Phoenix, AZ 85036
(602-244-6900)

National Semiconductor Corp.
2900 Semiconductor Drive
Santa Clara, CA 95051
(408-732-5000)

NEC Electronics Inc.
401 Ellis Street
Mountain View, CA 94039
(415-960-6000)

Precision Monolithics, Inc.
1500 Space Park Drive
Santa Clara, CA 95050
(408-246-9211)

Raytheon Co.
Semiconductor Division
350 Ellis St.
Mountainview, CA 94042
(415-968-9211)

RCA Solid State Division
Somerville, NJ 08876
(201-685-6000)

Signetics Corp.
811 E. Argues Avenue
Sunnyvale, CA 94086
(408-739-7700)

Standard Microsystems Corp.
35 Marcus Blvd.
Hauppauge, NY 11787
(516-273-3100)

Teledyne Semiconductor
1300 Terra Bella Avenue
Mountainview, CA 94043
(415-968-9241)

Texas Instruments, Inc.
P.O. Box 5012, Mail Station 84
Dallas, TX 75222
(214-238-2011)

Thomson Components - Mostek Corp.
1310 Electronics Drive
Carrollton, TX 75006
(214-466-6000)

Zilog, Inc.
1315 Dell Avenue
Campbell, CA 95008
(408-370-8000)

All addresses and telephone numbers listed in this section, while subject to change due to company activity, are correct as of the publication date.

III.5

Integrated Circuit Mechanical Data

Increases in integrated circuit component densities require that integrated circuit packaging technology also advance in order to meet the demand for additional pins and smaller physical outlines.

The following material illustrates the package formats used with modern integrated circuits. Figures III.5.1 through III.5.4 show the construction details for surface mount, pin grid array, and standard dual-in-line technologies. Figures III.5.5 through III.5.9 show the dimensional outlines for a variety of chip carriers.

Figure III.5.1 Plastic Chip Carrier
Available in 44-, 68-, and 84-pin configurations on 50-mil centers.
Courtesy of Gould Semiconductors

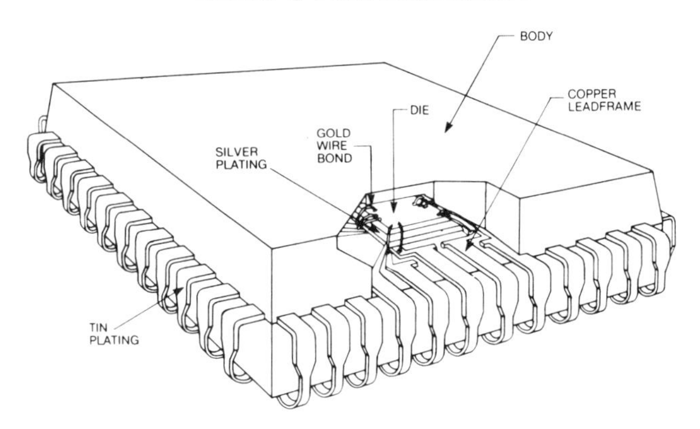

Figure III.5.2 Small Outline IC Package
Available in 16- and 28-pin configurations on 50-mil centers.
Courtesy of Gould Semiconductors

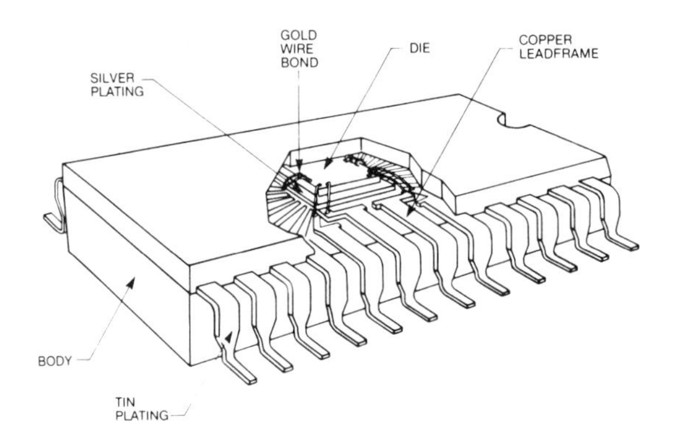

Figure III.5.3 Plastic Dual In-Line Package
Available in configurations of 8, 14, 16, 18, 20, 22, 24, 28, 40, 48, and 64 pins on 100-mil centers.
Courtesy of Gould Semiconductors

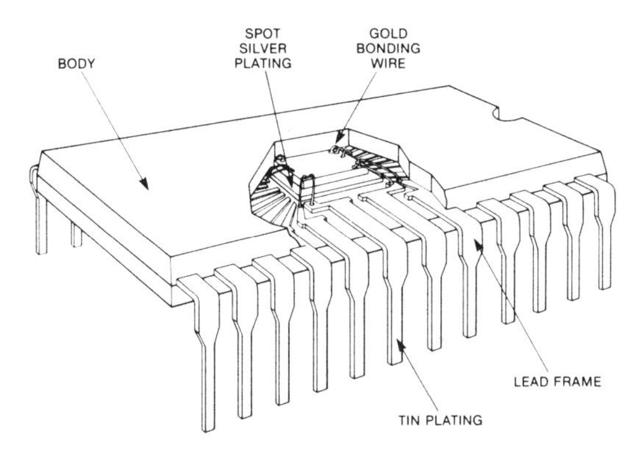

Figure III.5.4 Pin Grid Array
Available in configurations of 68, 84, 100, 120, and 144 pins.
Courtesy of Gould Semiconductors

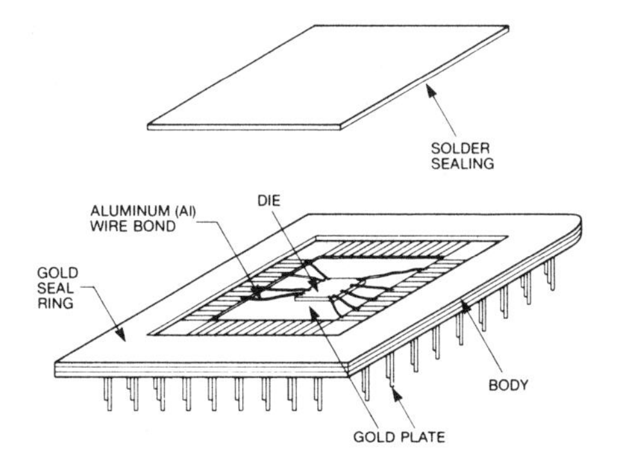

Figure III.5.5

Courtesy of Gould Semiconductors

28-Pin Cerdip

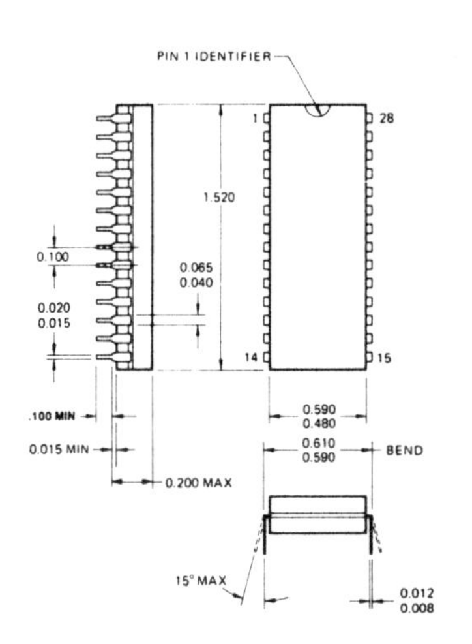

28-Lead Chip Carrier

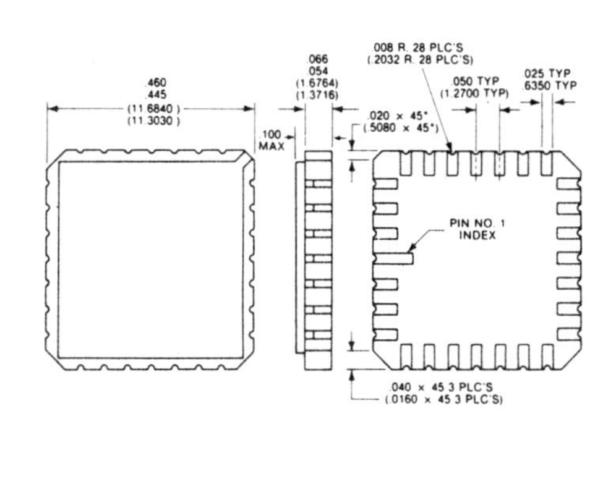

36-Lead Ceramic Chip Carrier

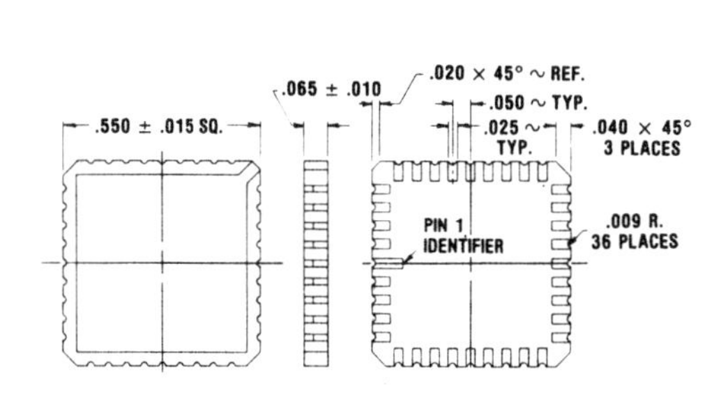

40-Pin Ceramic

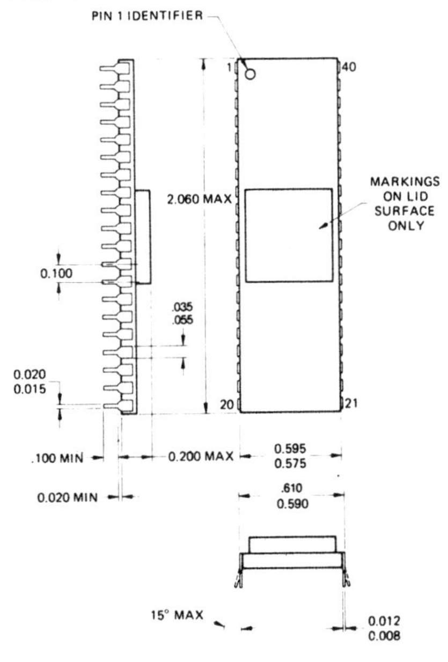

Figure III.5.6

Courtesy of Gould Semiconductors

24-Pin Plastic

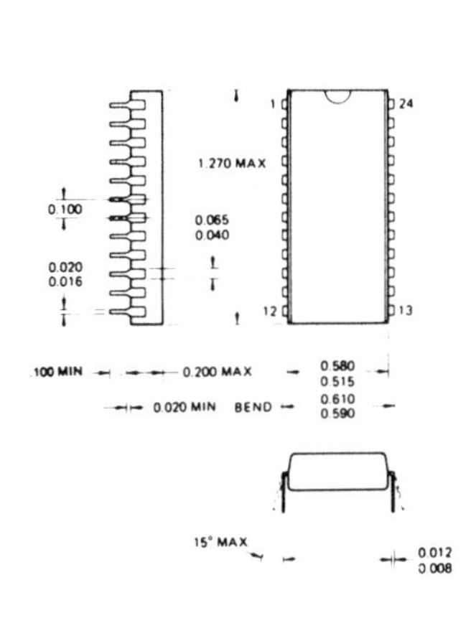

28-Pin Ceramic

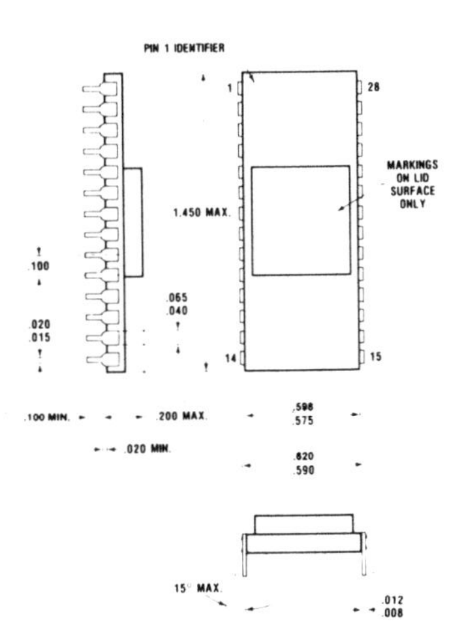

28-Pin Plastic

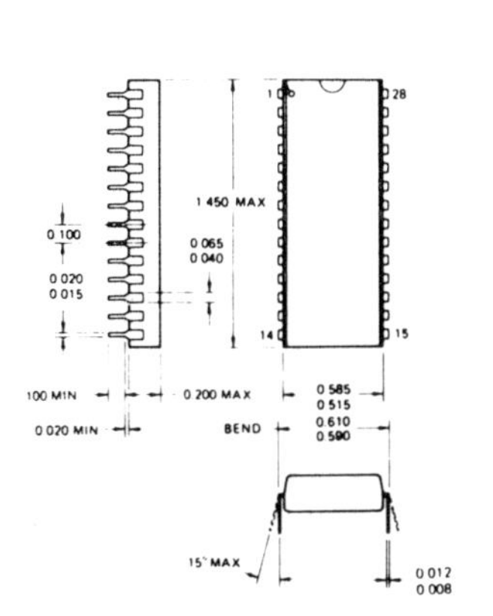

28-Lead Plastic S.O.I.C.

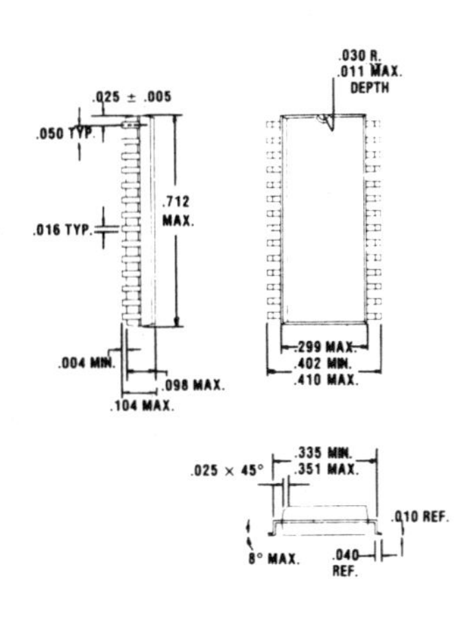

Figure III.5.7

Courtesy of Gould Semiconductors

40-Pin Plastic

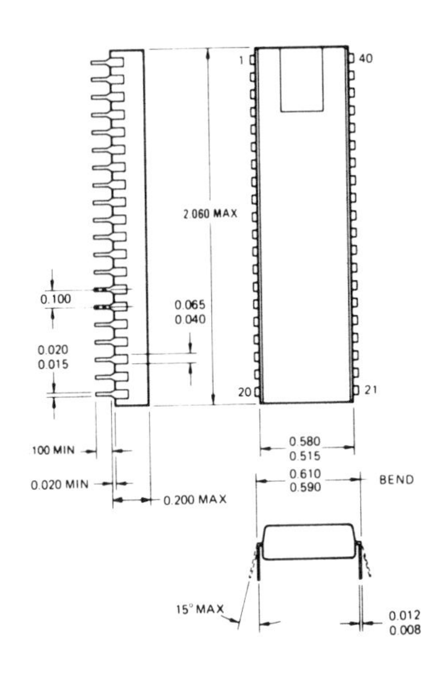

40-Pin Cerdip

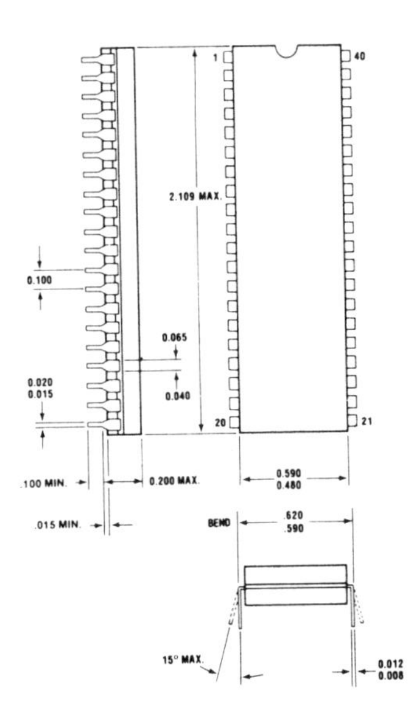

40-Lead Chip Carrier

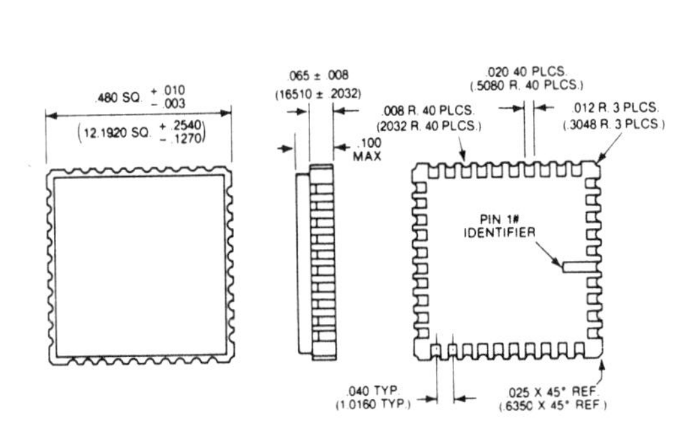

44-Lead Plastic Chip Carrier

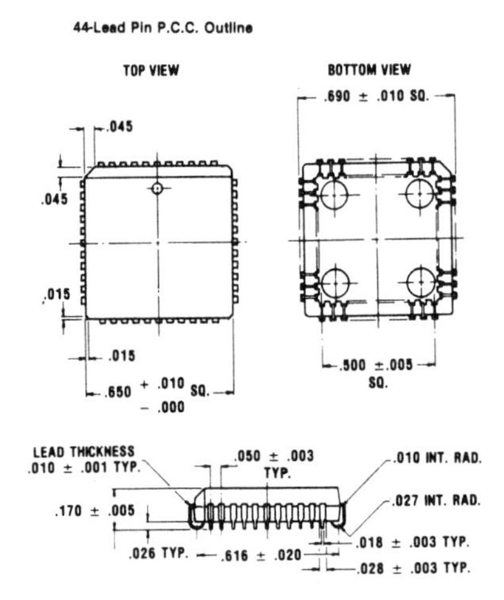

Figure III.5.8

Courtesy of Gould Semiconductors

48-Lead Ceramic

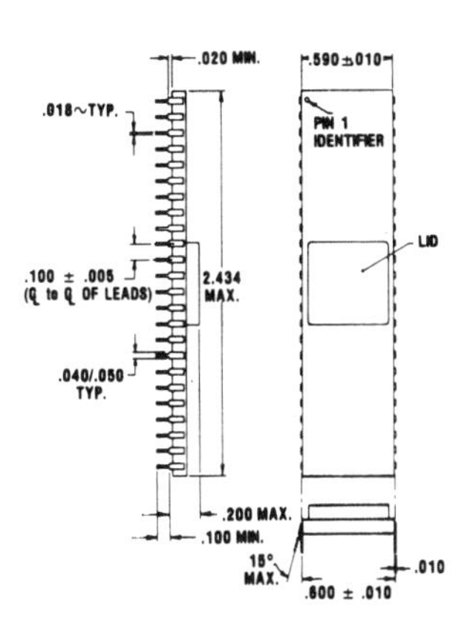

52-Lead Chip Carrier

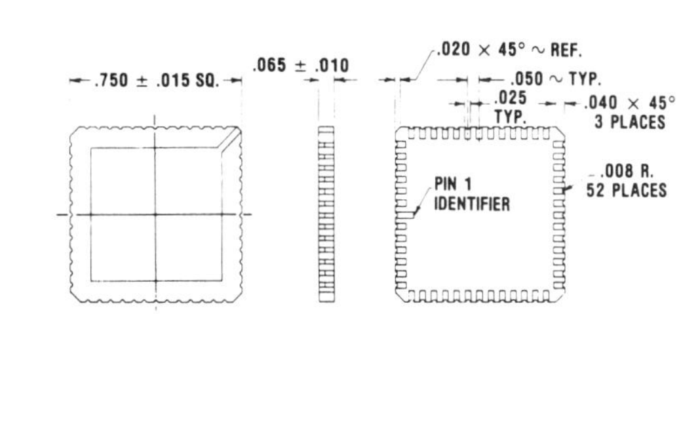

64-Pin Plastic

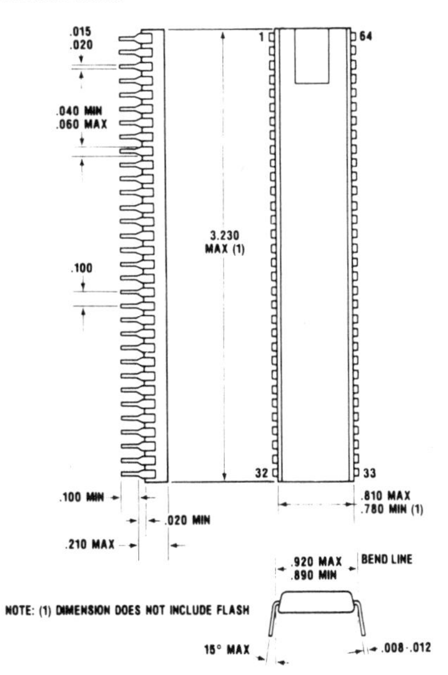

64-Pin Ceramic

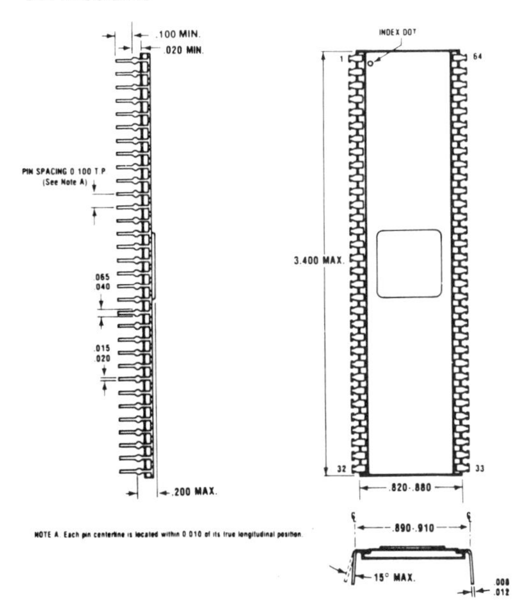

Figure III.5.9

Courtesy of Gould Semiconductors

68-Pin Grid Array

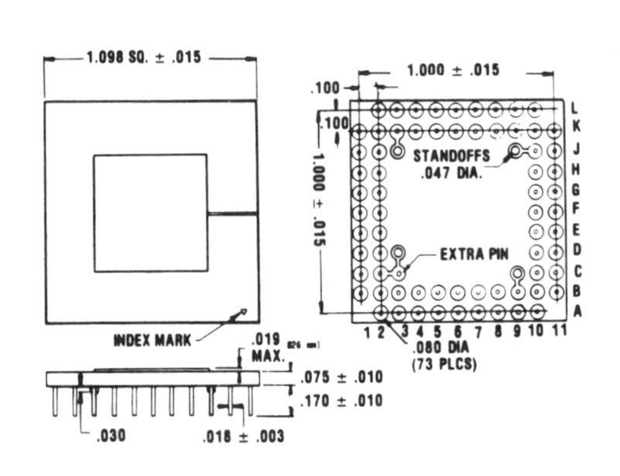

68-Lead Plastic Chip Carrier

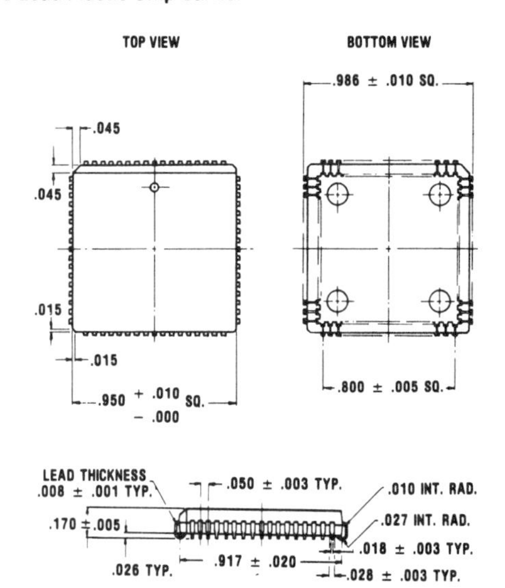

68-Lead Chip Carrier

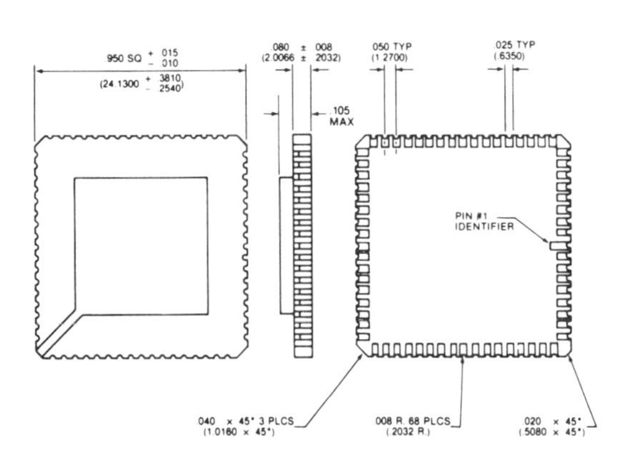

84-Pin Grid Array Outline

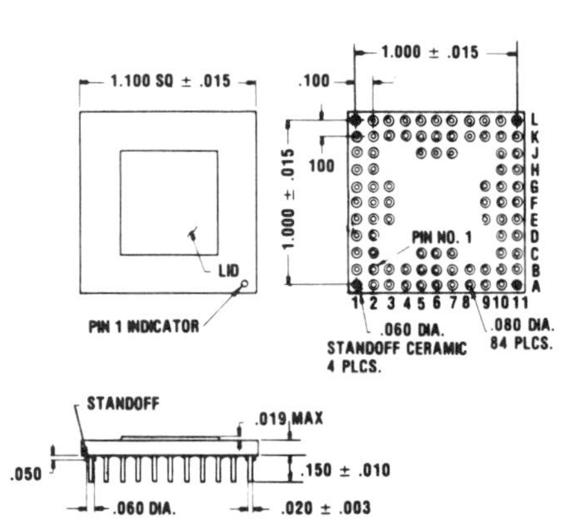

Index

A

B

C

D

E

K

L

M

R

S